Ecological HARMONY

V. SUNDARARAJU

INDIA • SINGAPORE • MALAYSIA

Notion Press

No.8, 3rd Cross Street
CIT Colony, Mylapore
Chennai, Tamil Nadu – 600004

First Published by Notion Press 2020
Copyright © V. Sundararaju 2020
All Rights Reserved.

ISBN 978-1-63606-952-4

This book has been published with all efforts taken to make the material error-free after the consent of the author. However, the author and the publisher do not assume and hereby disclaim any liability to any party for any loss, damage, or disruption caused by errors or omissions, whether such errors or omissions result from negligence, accident, or any other cause.

While every effort has been made to avoid any mistake or omission, this publication is being sold on the condition and understanding that neither the author nor the publishers or printers would be liable in any manner to any person by reason of any mistake or omission in this publication or for any action taken or omitted to be taken or advice rendered or accepted on the basis of this work. For any defect in printing or binding the publishers will be liable only to replace the defective copy by another copy of this work then available.

DEDICATION

This book is dedicated to all the Noble Warriors of the Tamil Nadu Forest Department who have sacrificed their precious lives in protecting the Forests and Wildlife.

CONTENTS

Foreword — 11

Preface — 13

Acknowledgements — 15

1. Protecting Heritage Trees Through A Special Law — 17
2. How Saving An Endangered Species Can Mitigate Climate Change? — 21
3. Caught In (Quick) Sand? — 25
4. How To Save A Coastal Dune? — 29
5. A Weed And A Tale — 33
6. Refugees In Their Own Home — 37
7. Timely And Prudent Action — 41
8. A Unique Dragon — 47
9. Tumakuru Sets A Green Example — 51
10. This Town In Karnataka Has Its Very Own Central Park — 55
11. To Create A Wild Haven — 59
12. Neelakurinji: Blue Bloom In A Blue Moon — 63
13. A Unique Tree Park For Chennai — 67
14. A Historic Judgement By The Supreme Court On Elephant Corridors — 71
15. E-Waste: How Should India Defuse A Ticking Time Bomb? — 77

16. How These Two Remote Districts In Andhra Are Tackling Wild Elephant Intrusions? 81
17. Green And Golden Srikakulam 85
18. Tamil Nadu's Innovative Methods Of Seed Plantation 89
19. How Alien Invasive Plant Species Threaten Western Ghats? 93
20. Amended Tamil Nadu Forest Act Must Be Cautiously Exercised 97
21. Why We Need A Coastal Zone Protection Act? 101
22. Raging Bull: How The Controversy Over Jallikattu Ended Up Saving The Pulikulam Breed 107
23. How To Save India's Elephants From Killer Rail Tracks 113
24. Implement The Biological Diversity Act In Its True Spirit 121
25. Eastern Ghats: A Biota Under Serious Threat 127
26. Propagate And Protect Halophytes 133
27. Scientific Management Of Mangroves Is The Need Of The Hour 139
28. Enact A Special Act For Protecting India's Natural Pharmacy 145
29. Cauvery: The River That The Tamils Thought Would Never Fail 149
30. Chennai Water Crisis: Some Suggestions To Avert A Redux 157
31. Why There Is An Algal Bloom In Gulf Of Mannar And How It Affects Livelihood 161
32. Why South India Needs The Shola Forests Of The Nilgiris 165
33. India Needs To Protect Its Wetland Flora 171
34. A Unique Institution For A Beloved Insect 175
35. Keeping The 'Green Hills' Truly Green 179
36. Tamil Nadu Must Sustainably Manage Its Rattans 183

37. India Must Protect Its Rare, Unique And Endangered Plants And Trees	187
38. Trichy's Elephant Rescue Centre Is A Haven For Abused Pachyderms	193
39. India Must Conserve Its Orchid Wealth	197
40. A Unique Rejuvenation Camp In Tamil Nadu For Captive Elephants	201
41. Why India Must Conserve Its Palm Trees	205
42. Suggested To Include The Watersheds In The H.A.D.P Map	211
43. Contribution Towards Conserving The Biodiversity Of The Nilgiris	217
44. Revival Of Shola Forests	223
45. Strategies For Increasing The Biodiversity Of Sigur Plateau In The Nilgiris	231
46. Role Of FPS In Conserving The Forests	243
47. Sleeping Giants Of Sivagangai	251
48. Fund Got Sanctioned By The N.B.M For Creating Shelterbelt	259
49. Conserving The Hills And Hillocks Outside The R.F	263
50. Saved Sakunthalai Malai	269
Works Consulted	*275*
Glossary	*279*
Abbreviations	*285*
About The Author	*295*

LIST OF BOXES

S.NO	TITLES	PAGE.NO
1.	What Is A Heritage Tree?	18
2.	Sea Grass And Its Eco-Services Sea Cow Or Dugong	22
3.	Watershed or Catchment Area	213
4.	Doddabetta Shola Nursery	224
5.	Revived Degraded Shola Forest, Made as A Model Trial Plot	227
6.	Rampant Smuggling of Sandalwood and Poaching of Elephants	233
7.	Crimsonbreasted Barbet or Coppersmith	237
8.	Rosewood-Vulnerable as per the IUCN Red List	247
9.	Virupatchi Forest Rest House	247
10.	500 Year Old African Baobab of Vedhiyarendal	251
11.	Baobab-Ancient Tree of India	252
12.	Baobab-a Magic and Sacred Tree	255
13.	Hills and Hillocks-Important Terrestrial Ecosystems	265
14.	Hill Area Conservation Authority (HACA)	266
15.	Ulakkai Aruvi Falls	270
16.	An Able and Devoted Person of the Kanyakumari Forest Division	272

FOREWORD

There is general consensus now, among all that life on planet earth is under serious threat due to gradual degradation and destruction of life-supporting systems. Among various elements of life-supporting systems, natural resources are high on our priority for special consideration as we ourselves are totally dependent on them for mere survival. Cosmos originated 15 billion years ago. Planet earth was formed about 4.5 billion years ago. Life evolved on planet earth 3.5 billion years ago and transformed into a human being only 2 million years ago. Hence it has taken billions of years for a perfectly balanced mega biosphere to be evolved with the capacity to support life on planet earth. None of us has, therefore any right to destroy the delicate balance of nature knowingly or unknowingly as it is meant for posterity. Look at the great irony of the situation we are in today. Given a choice, none of us wants to die. In fact, all of us are afraid of death. But all of us are killing ourselves knowingly or unknowingly by destroying the natural resources on which life is entirely dependent upon. Let us remember that all of us have come from nature and will have to return to nature one day. We must understand the intricate relationship betweenthe plants, animals and the habitat in which they live and conserve them with no compromise. The forests, hills, rivers, sea and sea coasts and mangroves are all under serious threat in our own country. The intricate relationship between the organisms and their relationship to the biophysical environment is called the ecology. In Harmonious existence, they all exist together and does no harm to each other but support each other.

Mr.V.Sundararaju.IFS (Retd), is a proud and knowledgeable forester with passion and dedication for the conservation of natural resources and particularly forests. He has worked tirelessly in various capacities in managing and protecting diverse forests in Tamil Nadu as Forest Range Officer, ACF, and DCF with zeal and enthusiasm and produced commendable results. He enjoys working in the field and does not stop his urge to learn more and more of conservation science.

This book on Ecological harmony speaks for itself the author's praiseworthy knowledge, experience and expertise on forest management. His flair for writing is clearly visible in all articles written by him in Down to Earth and compiled in this valuable book. He has also included his own experiences in managing the forest resources with certain case studies

It is a book worth reading by anyone interested in the conservation of Natural Resources and specifically forests. I wish him and his book "Ecological Harmony" the best.

C.K.Sreeedharan.IFS
Former Principal Chief Conservator of Forests and
Head of Forest Force (PCCF & HoFF), Chennai-600125.

PREFACE

Ecology is the study of interactions between organisms and their environment. We must maintain Ecological Harmony by protecting the environment and other living beings that inhabit the earth. If we maintain a harmonious relationship with each other, then only we can lead a happy life. With this idea in mind, I started writing scientific articles to Down to Earth website (www.downtoearth.org.in) on the advice of my esteemed friend Dr.R.K.Bharathi, former Chief Conservator of Forests at the beginning of 2018. Their spontaneous response motivated me to contribute more number of essays on different topics like Forests, Wildlife & Biodiversity, Environment, Water, Agriculture, Pollution, Mining, E-waste, etc. So far, the website has published 41 articles.

Heritage Trees, An Endangered Mammal that Mitigates Climate Change, Adverse Impacts of M-Sand, Coastal Dune, Unique Dragon, Coastal Regulation Zone Notification, Neelakurinji, Historic Judgement by the Supreme Court on Elephant Corridors, E-Waste, Alien Invasive Species of Western Ghats, Need of Coastal Zone Protection Act, Saving of Pulikulam Breed Through Jallikattu, Measures for Saving India's Elephants from Killer Rail Tracks, Eastern Ghats- a Biota Under Serious Threat, Protection and Propagation of Halophytes, Scientific Management of Mangroves, Need of Special Act to Protect India's Natural Pharmacy, Cauvery-the River that the Tamils Thought Would Never Fail, Suggestions to Avert Chennai Water Crisis, The Negative Impacts of Algal Bloom on the Livelihood of the Fishing Community, Importance of Shola Forests, Needs to Protect India's Wetland Flora, A Unique Institute for a Beloved Insect, Sustainable Management of Rattans, Orchid wealth of India, Necessity of conserving the palm trees and Needs to Protect the Rare, Unique and Endangered Plants and Trees are some of the interesting and thought provoking articles that every Nature Enthusiast should read and apply his mind in helping to implement the suggestions for fostering Ecological Harmony.

There is a lack of awareness about conserving natural resources even among the academic community. I have discussed in detail how the plants, wildlife, and different ecosystems in which they live are vandalised. I have pointed out the various urgent measures to be taken for protecting the same. As many may not have access to the internet, they might not have read the essays as the articles are published online. So, to make my articles related to various scientific subjects readily available for them to read and enjoy, I have brought out the same in a book form.

The later part of the book recounts some of the noteworthy happenings and exciting events that had taken place in my career in the Tamil Nadu Forest Department. And how best I had made use of the opportunities for sustainable management of Ecological Harmony have been documented in detail. Wherever I served, I had taken extraordinary interest and due care to conserve and improve the biodiversity of the region. I had initiated earnest efforts to protect the natural resources not only within the reserve forests but outside also. Hope this book may be of great use not only to the Foresters and Nature Enthusiasts but also to every individual who has great regard and affection towards Nature.

V.Sundararaju.IFS,
Former Deputy Conservator of Forests,
Tiruchirapalli-620017.

ACKNOWLEDGEMENTS

I am greatly indebted to Dr.C.K.Sreedharan. IFS, former Principal Chief Conservator of Forests and Head of Forest Force (PCCF & HoFF) for the valuable foreword offered by him.

I am happy in expressing my sincere thanks to Dr.R.K.Bharathi, former Chief Conservator of Forests for encouraging and guiding me to write to Down to Earth.

I express my gratitude to Thiru.K.R.Varatharajan.IFS, former Deputy Conservator of Forests for reviewing the manuscript and offering valuable comments and suggestions.

I am very much thankful to all the officers, field personnel and office staff of the Tamil Nadu Forest Department, especially to those with whom I had the great opportunity of serving for more than 36 years.

I wish to express my sincere thanks to my wife Johnsi Rani, who is a constant source of encouragement and guidance.

My gratitude goes to my sons Vimal and Praveen Kumar, daughter-in-laws Sumana and Jeeva and our lovely grand-daughter Viena for their continuous motivation and wholehearted support for publishing the book.

I am very much grateful to Notion Press, Chennai for publishing the book 'Ecological Harmony'.

V.Sundararaju.IFS,
Former Deputy Conservator of Forests,
Tiruchirapalli-620017.

1
PROTECTING HERITAGE TREES THROUGH A SPECIAL LAW

We have identified 100 heritage trees through the Society for Conservation of Nature (SOFCON) in Tamil Nadu and want the State Government to enact the Heritage Tree Conservation Act (HTCA) to protect these trees which provide invaluable ecosystem services.

In 2010, when I was serving in Kanyakumari district as the District Forest Officer, I identified a 500-year-old heritage tree called, Terminalia Arjuna tree. A heritage tree is a large tree whose value is considered irreplaceable. These trees take decades and centuries to mature and act as prominent landmarks of a place (see box).

Rapid development in the state has posed a great danger to these trees which provide invaluable ecosystem services such as shelter for wildlife, carbon sequestration, release of oxygen, shade, soil conservation, creation of microclimate, eco-tourism, etc. Thus, the society has been highlighting the issue of conserving and protecting heritage trees through talks in educational institutions and other public forums such as Rotary and Lions clubs in the state.

What is a Heritage Tree?

- The main criteria for considering a tree as heritage tree is its size, form, shape, age, colour and rarity.
- The aesthetic, botanical, horticultural, ecological, social, cultural and historical values are also taken into account.
- A specimen associated with a historic person, place, event or period is also treated as heritage tree.
- A heritage tree can also be a tree associated with local folklore, myths, legends or traditions.
- Certain species that are relatively rare in an area, whether native or not, may also be granted special status.

Real solutions

We are also suggesting to the State Government to enact a special law, the Heritage Tree Conservation Act (HTCA) to impose restriction upon removal of these precious trees.

A Heritage Tree Conservation Committee (HTCC) can be established in every district to conserve heritage trees. The Chairman of the committee may be the District Collector and other members may be the District Forest Officer (DFO), Joint Director (JD) of Agriculture, Joint Commissioner (JC) of (HR & CE), Joint Director of Horticulture, Divisional Engineer (DE) of (PWD), Revenue Divisional Officer (RDO), Tahsildar, one Scientist, one non-profit and one social activist.

Apart from meeting periodically and inspecting the trees, the committee can help create awareness among the public to increase the tree cover throughout the state. This is because there is no scope to expand the forest cover due to non-availability of land.

As most of the heritage trees in the state are found in temple premises, the Hindu Religious and Charitable Endowment (HR & CE) department authorities can be trained about how to protect the trees. Health cards can also be prescribed for the identified trees and they can also be used to take action to remove any dangerous trees.

When the heritage trees are identified in private lands, land owners can be honoured with suitable rewards in order to motivate them. Tree cover expansion can be achieved only with the wholehearted involvement and support of the land holders especially in the rural areas. Unless we have one third of the land area under forest and tree cover, we can't attain economic stability.

Photo by Author

A 500-year-old tree called Terminalia Arjuna-Neer Maruthu (Tamil), Thella Maddi (Telugu), Nirmathi (Kannada) and Arjun (Hindi) in Kanyakumari district, Tamil Nadu State

http://www.downtoearth.org.in/blog/protecting-heritage-trees-through-a-special-law-59703. Thursday 15 February 2018

2
HOW SAVING AN ENDANGERED SPECIES CAN MITIGATE CLIMATE CHANGE?

The Thanjavur Forest Division has been conserving sea grass in Palk Bay in order to save the endangered sea cow as well as mitigate climate change.

Dugongs are marine mammals that relish sea grass, the most productive plant communities. Because dugongs and sea grass species (see box) are interdependent and interrelated, if dugongs become extinct locally, sea grass meadows may gradually disappear. Sea grass is an important resource also because it sequesters up to 11 per cent of the organic carbon buried in the ocean even though it occupies only 0.1 per cent of the total ocean floor. Conserving dugong, thus, improves not only sea grass but also helps mitigate global warming.

Sea grass grows in abundance in Palk Bay, a strait between Tamil Nadu and the Mannar, Northern Province, Sri Lanka and a highly productive coastline in the southeast coast of India. The Tamil Nadu Forest Department has taken up a sea grass rehabilitation project which also targets the conservation of dugongs near Manora village, Thanjavur, Tamil Nadu in Palk Bay.

The Species Conservation Action Plan for Sea Cow was organised by the Thanjavur Forest Division in 2016 and 2017 under the Tamil Nadu Biodiversity Conservation and Greening Project (TBGP). The Japan

International Cooperative Agency (JICA) financially supported the project and various departments were involved in the training workshops, including marine police, fisheries, veterinary and police.

Sea grass and its eco-services	Sea Cow or Dugong
Sea grass absorbs carbon from the atmosphere (up to 83 million metric tons of carbon annually) and also absorbs carbon from water to build their leaves and roots.	Found along the coasts of about 48 countries, dugong is a "Critically Endangered Species" as per the International Union for Conservation of Nature.
When parts of sea grass plants and associated organisms die and decay, they get buried in the sediment.	There are just 250 dugongs in India spread across Gulf of Mannar, Palk Bay, Gulf of Kutch and Andaman & Nicobar Islands.
One acre of sea grass can sequester 740 pounds of carbon per annum, the same amount emitted by a car travelling around 6,212 km.	Dugong is the only marine herbivorous mammal.
The complex architecture stabilises the bottom sediments, which reduce wave energy and current velocity. Since they reduce turbidity, coastal erosion is decreased.	Dugong seek shelter in sea grass and are inter-dependable.

Sea grass meadow in Palk Bay serve as a breeding and feeding ground for fish, molluscs, invertebrate species and mammals including, Dugong (Sea cow). Sea horses, sea cucumbers and pipe fishes are other co-habitants.

Due to the threats faced by dugong from humans, crocodiles, large sharks and killer whales, their former distribution range in certain parts of the world is now absent. Research has shown that where there is no human impact, dugong population increases only by about 5 per cent per annum. If more than about 2 per cent of adult female dugongs are killed every year, their population will decline drastically. Dugongs are harvested for food, meat, oil, medicaments, etc. When females are hunted, it leads to reduction in the breeding stock.

Activities such as pollution, trawling and silt accumulation by mining, mismanagement of catchment or coastal development has an adverse impact on the population of dugongs. Loss of sea grass due to large scale floods can destroy their feeding and breeding grounds. The noise by vehicles such as boats may scare the animals and fishing lines and nets can prove fatal.

The endangered species can be saved

Current and long-term monitoring of dugongs shows that their populations can be maintained or recovered by ensuring protection of their habitats, reducing their deaths due to fishing. Research and monitoring scientists are tracking dugongs through the aerial survey method to determine the grazing areas, duration and depths of dives, movements between grazing areas and between regions. By identifying the main feeding areas through aerial tracking, the management of net fishing and boat traffic in these areas are regulated. Population management can be done by creating awareness among the fishing community.

Fishermen take charge

Awareness programs have been organised in many coastal villages such as, Kazhumanguda, Karanguda, Mallipattinam, Chinnamanai, Manora, Velivayal, Pillayarthidal, Somanathanpattinam and Sethubhavachathiram along the coast of Thanjavur. Street plays with dance, music and drama explained the value of sea grass for sustainable fishing and conservation of dugongs. Hoardings, booklets and brochures were distributed among the fishing villages, schools, colleges and other line departments.

Fishermen were motivated to release dugongs into the sea by giving awards. Since releasing dugongs means the expensive nets have to be cut open, they were given compensation by the forest department. The awareness campaigns proved meaningful as the authorities released dugongs caught accidentally by fishermen, twice. On December, 2016, the fishermen from Keezhathottam village who caught two dugongs accidentally informed the Forest Range Officer (FRO), Pattukottai. The dugongs were released into the sea at Manamelkudi near Kattumavadi. One more dugong was also released at Kodimunai on

January 30, 2017 and the fishing villagers who released the dugongs were duly compensated by the Forest Department. Besides, the persons who gave information about the movements of dugong were also rewarded.

Dugong-Seacow

Courtesy: Google

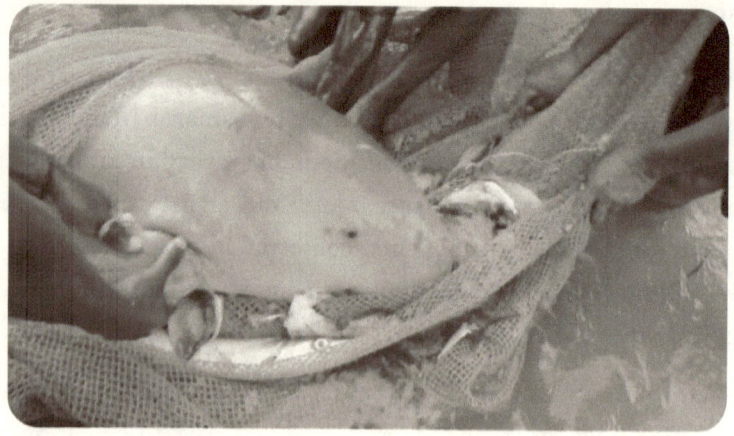

Courtesy: Pattukottai Forest Range

Releasing of Dugong by fishermen cutting open the fishing net

http://www.downtoearth.org.in/blog/how-saving-an-endangered-species-can-mitigate-climate-change-59867

Monday 12 March 2018

3
CAUGHT IN (QUICK) SAND?

The Tamil Nadu Chief Minister's statement in January, saying the state would officially promote 'M-sand' as a substitute for river sand is uncalled for.

In early January, Tamil Nadu Chief Minister was speaking in the state Assembly on the shortage of sand, used extensively in the construction sector. Tamil Nadu has been in the throes of a sand shortage since last year. Experts say the crisis is primarily due to the lapse in many sand quarry leases and the increased smuggling of sand to neighbouring states. Also, late last year, the Madras High Court ordered the closure of sand mines for about six months and stayed the opening of new sites.

The shortage of sand, which India categorises as a "minor mineral", has become a political hot potato in Tamil Nadu. The Chief Minister was replying to Dravida Munnetra Kazhagam (DMK) MLA Thangam Thennarasu, who wanted the state government to bring down the price of sand. The chief minister said, "The time has come for us to switch over to M-sand. We may have no option but to use M-sand in the days to come because sand quarries will be closed in three years," he told the House. Tamil Nadu's Minister for Law, Mines and Minerals informed that the government would promote entrepreneurs opting for M-sand production by facilitating easy credits and power subsidy.

Let us first understand what "M-sand" is. It is produced by crushing hard granite. When big size boulders are put into giant size machines, they are broken into "40 mm metal" first and then into 20 mm metal. On further processing, they are crushed to 4.75 mm, 3.5 mm and 2.36 mm granules called M-sand. The crushed sand is of cubical shape and with grounded edges. It is washed and graded as a construction material.

Coming back to the topic at hand, the Chief Minister's announcement appears absurd. How will the government manufacture M-sand? Boulders are required for that. From where will they get the boulders? They will have to destroy the state's hills and hillocks for that purpose. How many hills and hillocks are there in the state? How can they ensure the sustainable manufacture of M-sand? Before making its announcement, the government should have thought of all these things. It seems to have been made just to convince builders and consumers, without applying the mind.

Hills and hillocks are a very important terrestrial ecosystem. They support a variety of plants and animals (including micro-organisms) with grasses, herbs, shrubs, climbers, creepers and trees that have grown on them. They absorb carbon dioxide and release oxygen, conserve soil and water, in addition to moderating the climate of the particular region.

In the past, hills and hillocks in Tamil Nadu, as in the rest of India, had been usually left untouched given that they were considered to be the abode of religious deities, gurus, monks and seers. This is no longer the case. Today, many hillocks have been vandalized by greedy quarry contractors in districts like Madurai and Kanyakumari. Certain hillocks have been notified by the Revenue authorities as "water bodies" in districts like Madurai. For example, the Pancha Pandavar hillock, an archaeological site, has been listed in the revenue records of Madurai district as a water tank in the Keezhavalavu panchayat. Likewise, as the hillock with Jain stone beds dated back to the 2nd century BC has not been entered in the Prohibitory Order Book, it has been quarried, even after repeated representations by the President of the Keezhavalavu Panchayat to Collectors.

If the manufacture of M-sand is permitted, within no time, the state's hillocks would disappear. Then, the manufacturers may gradually encroach upon the hills of the Eastern and Western Ghats. They would follow this by trying to swallow the hills of other protected areas in the name of development, thereby depriving the people of the state, pure air, quality water, fertile soil, food, shelter, etc. The attractive slogan of "eco friendly and economical alternative for river sand" floated by M-sand manufacturers is actually a bogus one. The people of Tamil Nadu must put up a vigilant fight against M-sand as the manufacturing of the same may cause irreparable damage not only to our ecosystems but also to our basic survival.

That does not mean, however, that river sand mining should be allowed indiscriminately. Transparency is needed while permitting it. Mining should be done only by the Public Works Department (PWD). Machinery can be used but the depth should be strictly followed (not below 1 metre). The Environmental Impact Assessment Committee (EIAC), before submitting any report about the inspection of the sand mine to the government, should conduct a public hearing with nearby village residents and conservation-oriented NGOs. The Committee Members must be able to convince the people and remove the suspicion from their minds. Only after that, the report should be submitted. Wherever there are permitted quarries, there should be a hoarding displayed in a prominent place clearly mentioning the extent, period of quarrying, quantity permitted, etc. The quarry site should be clearly demarcated on the ground with semi-permanent structures. The online payment for sand with the PWD may be made compulsory.

Tamil Nadu's current scarcity of sand seems to be artificial. The PWD should regulate sand mining and take up the responsibility of protecting rivers. Periodic checks by the Vigilance and Anti-Corruption wing of the State Police Department and a team of experts as ordered by the National Green Tribunal (NGT) may reduce many problems. The export of sand to other states and countries should be strictly banned. No more fresh permissions should be granted for establishing new crusher units in the state. The crusher waste can be processed further for using as an alternative for sand with the scientific inputs given by institutes like the Indian Institute of Technology (IIT).

Courtesy: Google

Destroying Hillock for manufacturing 'M sand'

http://www.downtoearth.org.in/blog/caught-in-quick-sand-601521

Wednesday 11 April 2018

4

HOW TO SAVE A COASTAL DUNE?

The struggle of one village at India's southern tip to save itself from corporate interests is a tale in itself.

The district of Kanyakumari in Tamil Nadu is situated at the southernmost tip of the Indian Sub-continent, surrounded by the Bay of Bengal on the east, Indian Ocean on the south and the Arabian Sea on the west. It is famous as a pilgrimage and tourism spot.

In the district, Rajakkamangalam beach has the Arabian Sea on its west and abuts the village of Pannaiyoor, in the Agastheeswaram *taluk*, on its east. There is a sand dune formed naturally on the shore. Pannaiyoor owes its very existence to this coastal sand dune. The village has been well protected from any kind of natural disaster because of it.

This is because coastal sand dunes act as a natural barrier against wind and waves and protect inland areas from damage due to storms, cyclones and tsunamis. Sand dunes are also an important ecosystem and support highly specialised plants and rare and endangered animal species. Naturally occurring tree species like *Calophyllum inophyllum*, *Borassus flabellifer* and *Pandanus fascicularis* are found on coastal dunes in addition to herbs like *Acalypha indica*, *Aloe vera*, *Argemone maxicana*, *Calotropis gigantea*, *Leucas aspera*, *Gisekia pharnaceoides*, *Ipomoea pescaprae* and *Tephrosia purpurea*. The dunes are stabilised by the vegetation that grows on them, binding the sand particles.

Sand dunes up to a height of about 20 metres are found along the Tirunelveli, Kanyakumari and Kancheepuram coasts. Important sand dunes in Tamil Nadu are found in Tiruchendur, Manapad, Ovari and Idinthakarai of Tuticorin district. In Kanyakumari district, they are found in Kanyakumari, Chothavilai, Pallamthurai, Sankuthurai, Periakadu, Chinnavilai and Rajakkamangalam.

However, in spite of being very important for the survival of Pannaiyoor, the dune at Rajakkamangalam was about to be obliterated. The Tamil Nadu Coastal Zone Management Authority, in its recommendation dated June 29, 2010, called for the development of a fishing harbour in Agastheeswaram *taluk*. According to the recommendation, the project site was in Coastal Regulation Zone (CRZ)-I (ii) and CRZ-III. Based on the recommendation, the proposal was considered by the Expert Appraisal Committee and recommended for grant of CRZ clearance with certain conditions. Accordingly, on July 18, 2011, the Union Ministry of Environment, Forest and Climate Change (MoEF&CC) granted CRZ clearance to M/S Rajakkamangalam Thurai Fishing Harbour Pvt Ltd for the development of the fishing harbour.

On hearing that the path had been cleared for the process of establishing a private harbour in their vicinity, the residents of Pannaiyoor worried that any damage caused to the natural sand dune might bring a catastrophe upon the entire village. They tried their level best to stall the project through the Citizens Welfare Trust formed in the village as their very survival was at stake. But they were made to run from pillar to post as they did not have either money or any other influence. When it was confirmed that order had been issued in favour of the private fishing harbour, they had no other alternative except to knock on the door of the highest authority.

On the representation from the Pannaiyoor Region Citizens Welfare Trust and the Conservation of Natural Trust (a renowned NGO known for protecting the environment of the district) alleging the project was in the sand dune area, a site inspection was conducted by the Additional Principal Chief Conservator of Forests and the Scientist 'F' Regional Office, MoEF&CC, Chennai on January 17, 2015. The site inspection revealed that the project area had sand dunes and hatcheries of sea turtles. But as per the 2011 CRZ Notification, the sand dunes are classified as CRZ-I (B) where most activities are prohibited except some like laying of conveying system and projects relating to the Department of Atomic Energy.

On receipt of the inspection report, the MoEF&CC understood that the recommendation of the state government authorities allowing the construction of a private harbour lacked correctness with respect to the presence of sand dunes at the project site. The Union Government, exercising the power conferred on the Ministry under Section 5 of the Environment (Protection) Act, 1986, cancelled the CRZ clearance and thus the valuable sand dune at Rajakkamangalam was saved.

Courtesy: Google

A coastal sand dune that acts as a barrier against cyclone & tsunami

http://www.downtoearth.org.in/blog/how-to-save-a-coastal-sand-dune-60390

Wednesday 02 May 2018

5

A WEED AND A TALE

In the forests of Kanyakumari, an invasive species has been proliferating since the 1960s. It is time to remove it altogether.

In the 1960s, ethnic strife in the island nation of Sri Lanka, which ended in 2009, was rife. It caused many Sri Lankan Tamils to cross the Palk Strait and enter Tamil Nadu, seeking refuge. India's then- Prime Minister, Lal Bahadur Shastri and the then-Lankan President, Sirimavo Bandaranaike signed an agreement to accommodate the refugees. As most of them had been working in plantations like tea and rubber in Sri Lanka, it was decided to create similar plantations in Tamil Nadu to employ and rehabilitate them. Accordingly, forests in the districts of Kanyakumari and Nilgiris were chosen and converted into rubber and tea plantations.

The Kanyakumari Forest Division leased about 5,000 hectares of reserve forests to the Arasu Rubber Corporation (ARC) Limited for raising rubber plantations to rehabilitate Tamil refugees from Sri Lanka. Initially, when rubber was raised, *Pueraria javanica* was introduced as a "cover crop" to prevent soil erosion and for enrichment of the soil. This species is native to Thailand. It is a very popular cover crop in coffee, oil palm, citrus and rubber plantations.

Subsequently, the cover crop was changed. During the 1980s, *Mucuna bracteata*, a creeping vine that belongs to the legume family, was introduced as cover crop in place of *Pueraria javanica*. *Mucuna bracteata* originates in the forests of Tripura. It is used as a cover crop in rubber (India and Malaysia) as well as palm oil plantations (Malaysia).

Mucuna bracteata has several advantages. The creeping crop grows about 10-15 cm/day in rubber and palm oil plantations. It mainly protects raked up soil, prevents erosion, enriches organic matter and helps in retaining moisture. It minimises losses of nutrients due to leaching and reduces competition from noxious weeds. As a rule, leguminous plants are selected as cover crops as they are able to fix nitrogen and make it available to the main crop. *Mucuna bracteata* does not dry during hot spells. It also grows under the shade. It is non-palatable to cattle due to high levels of phenolic compounds and thus protects crops from cattle and other wild animals. Moreover, it is drought-resistant and poses less fire hazards during dry weather.

Unfortunately, *Mucuna bracteata* is also a deep-rooted and rapidly-growing plant species. In the ARC plantations, the growth of this species became astonishingly vigorous in due course of time, and has now become a menace in the adjoining forest areas.

Since it is a fast-growing, creeping and aggressively climbing, perennial vine, it can choke, smother and dwarf native trees by its gregarious growth and climbing behaviour. It has covered the ground completely, climbed upon the well-grown trees and suppressed them completely. Because of the suppression by the weed, the photosynthetic activity has completely stopped. As a result, the trees are gradually facing death.

The natural forest adjoining the ARC's plantations needs revitalisation. Moreover, it is rich elephant habitat. Therefore, it is the need of the hour to remove this weed completely from the ARC areas contiguous to the forest as this poses a serious threat to the wildlife habitat which has been declared as a Wildlife Sanctuary. In the past, attempts have been made with machinery like JCBs. However, the weed could not be uprooted completely. Even if a small bit of its root is left in the soil, it will grow vigorously and cover the whole area within no time. It is very difficult to remove the weed once it is established.

As *Mucuna bracteata* has become a major threat, it is high time that the Forest Department and the ARC jointly embark on a programme to eliminate the invasive species for better management of forests. Legislators, foresters, scientists and environmentalists should come together for enacting a strict legislation to prevent the use of this harmful weed in plantations nearer to natural forests.

Photo by Author

Abandoned ARC rubber plantation and the well-grown trees adjoining the boundary fully covered by Mucuna bracteata, a menaceful invasive species

http://www.downtoearth.org.in/blog/a-weed-and-a-tale-60559

Wednesday 16 May 2018

6
REFUGEES IN THEIR OWN HOME

In Kanyakumari Wildlife Sanctuary, thousands of hectares of land leased to private contractors is threatening wildlife in the region

In the heart of Kanyakumari Wildlife Sanctuary, an ill-thought move by a rubber plantation corporation is exacerbating human-animal conflict. Crops such as pineapple, banana and tapioca are being cultivated in thousands of hectares of Arasu Rubber Corporation (ARC)'s young rubber plantations. Since the wild animals such as the Indian bison, sambar and elephant see this as food, not only do the crops get damaged, it also leads to human-animal conflict.

The authorities concerned say the plantations are leased out to private contractors only for three years and helps avoid the growth of unnecessary weeds. They add that this has been done on the suggestion of the Rubber Research Institute of India with enough security measures like solar power fencing and elephant proof trenches. However, this is ultimately posing hardships to the wild animals in their own habitat.

Crop raiding is a major problem in areas when agricultural crops are grown adjoining forest areas. The contractors protect their crops from the wild animals by using crackers and throwing lighted coconut shells packed with ash and dipped in kerosene oil. Animals, especially the elephants, are likely to get injured and the pachyderms in turn may become more aggressive leading to human fatalities.

Presently, the Arasu Rubber Corporation covers five divisions, namely Keeriparai, Manalodai, Kodayar, Mylar and Chithar. In Keeriparai and Manalodai, where almost 1,000 hectares have been leased out to private contractors for raising pine apple, banana and tapioca, several animals such as elephants, the Indian bison, sambar (Kada Maan), barking deer (Kelai Aadu), mouse deer (Sarugu Maan), sloth bear, wild boar and porcupine are found.

Environmentalists, naturalists and other social activists from Kanyakumari district say that the practice of leasing out the young rubber plantations to private contractors for raising palatable crops like pine apple, banana and tapioca should be stopped in order to provide an undisturbed habitat for the wildlife.

The entire Kanyakumari Forest Division was recognised as a Wildlife Sanctuary in 2002 for better protection of wildlife and their habitats. Apart from ARC, there are two corporations under the Tamil Nadu Government-Tamil Nadu Tea Plantation Corporation (TANTEA) and the Tamil Nadu Forest Plantation Corporation (TAFCORN). While the two corporations were started with the idea of rehabilitating the Sri Lankan refugees in 1960, TAFCORN was started to provide the pulpwood to wood-based industries in the state. There is no violation of Wildlife Protection Act.

The farmers outside the reserve forest areas are given compensation for crop damage caused by wildlife in their fields after due assessment made by the Agriculture, Horticulture and Forest Officials. In case of any human death or injury caused by wild animals, due compensation is paid by the Forest department. In addition to all these, erection of solar power fence besides digging of elephant proof trenches are also carried out around the reserve forests wherever elephant movements are noticed.

Already the forests of Kanyakumari have been fragmented badly by large number of private forests, irrigation dams, leases to different departments, encroachments and tribal settlements. Of the 11 dams in the district, 9 are located within the sanctuary. There are major encroachments like Valiya Ela that took place when Kanyakumari district was bifurcated from Kerala state and attached to Tamil Nadu state. In addition, there are many more encroachments against which umpteen numbers of cases are pending in various courts. There are about 47 tribal settlements with agricultural lands well within the sanctuary limit.

If the Arasu Rubber Corporation is particular about avoiding weed growth and providing additional income, the lessees can be allowed to raise species such as citrus, chilli, capsicum, sesame, medicinal plants, which are not relished by wild animals.

Photo by Author

Thousands of hectares of land in rubber plantations run by state corporation are leased to contractors who grow palatable crops. This threatens wild animals in the area such as sambar and Indian bison.

http://www.downtoearth.org.in/blog/refugees-in-their-home-60680

Tuesday 29 May 2018

7

TIMELY AND PRUDENT ACTION

That is what saved a part of Kanyakumari town's coastal strip during my tenure about 8 years ago.

It was November 2010. I was serving as the District Forest Officer (DFO) at Kanyakumari, Tamil Nadu. During my tenure, an incident took place which reaffirmed the townsfolk's faith in the administrative system that is used to govern us.

The matter had its origins in 2007. In an order dated August 17 of that year, the Executive Officer of the Kanyakumari Special Grade Town Panchayat had granted licence to a hotelier to maintain the Beach Road Park, located next to the Kamaraj Memorial along the Beach Road in Kanyakumari, for a period of three years from 2007 to 2010 on an annual lease of Rs 16,000. The order was granted following the tender formalities and the passing of a resolution by the Town Panchayat.

Based on these procedures, the Kanyakumari District Collector (DC) granted permission in his order dated September 20, 2007 to the Town Panchayat to entrust the beautification and maintenance work of Beach Road Park located opposite to Tamil Annai Park and Sunset Town Park.

Subsequent to the order given by the authorities, the hotelier started demolishing the existing cement structure of the Beach Road Park located opposite to Tamil Annai Park with the idea of constructing a new structure. Since I was known to many in the district, sometime during 2008, some social activists and environmentalists informed me about the demolition work which was going on. Finally, one day evening, I visited the site. A JCB was carrying out demolition work. When I enquired the driver, he said the work of improving the park by re-modelling the existing structure had been assigned to a hotelier and as per his instruction, the demolition was being carried out. I did not say anything and returned quietly.

Subsequently, I informed the DC about the complaints received and my inspection of the spot. Meanwhile, two organizations, namely "Indian National Trust for Art and Cultural Heritage" (INTACH), and the "Sunset Bazaar Viyabarikal Nala Sangam" (SBVNS) filed a Writ Petition before the Madurai Bench of the Madras High Court contending that the area in question came under Coastal Regulation Zone-I (CRZ-I) and neither any construction nor any developmental activity was permissible as it would destroy the environment and ecology of the beach (W.P (MD) Nos.4422 and 5306 of 2008). Based on the Writ Petition, an interim direction was issued by the said court sometime during May, 2008, to maintain status quo in addition to ordering a notice. Subsequently, both, the District Collector and the Executive Officer (EO) revoked the licence granted.

The DC and petitioners claimed in the public interest litigation that the area in question was in CRZ-I. While the Licensee and EO claimed that the license granted was only for the maintenance of the existing park, the petitioners claimed that there was no park at all. The DC had observed that in violation of the conditions of the licence, the Licensee was trying to raise a structure after demolishing the existing structure. This was refuted by the Licensee saying that he had only attempted to renovate and maintain the park. While the Licensee claimed that the establishment and maintenance of a park or playfield is a permitted activity under the notification issued by the Union Ministry of Environment, Forest and Climate Change requiring no clearance, the DC and petitioners said that even for that, the permission of the District Coastal Zone Management Committee (DCZMC) should have been obtained.

Considering these facts carefully, the Court directed the DC to convene a meeting of the District Coastal Zone Management Committee (DCZMC) within a period of three weeks, place the entire records before the DCZMC and make it inspect the place in presence of the Licensee. After inspection, the DCZMC was to probe: a) Whether the area fell under CRZ-I, II or III? b) Whether there was a park in existence? c) Whether the activities carried out were in tune with the law? d) Whether any action was needed relating to preservation of the environment? e) Whether any clearance was required for the establishment of a park?

The Court closed the Writ Petition directing the DC to complete all the proceedings within a period of three months besides allowing the Licensee to carry on only routine maintenance work without altering the character and features of the land.

In order to convene a meeting and pass necessary orders in this matter, the DC wanted to inspect the area in question. He had sent me a letter, requesting me to attend the inspection as the DFO is a member of the DCZMC. On that particular day, in November, 2010, as I had to conduct a written test for the selection of forest guards, I sent my Assistant Conservator of Forests (ACF) to Kanyakumari beach. At about 10 o'clock, the DC called me over phone, requesting me to come over personally, as the matter was related to a case pending with the Madurai Bench of Madras High Court. I went to the spot and met the DC. Then, the inspection of the said park in question was taken up. When we went to the seashore, I asked for the map with the High Tide Line (HTL) and Low Tide Line (LTL) marked. The officials of the Local Development Authority (LDA), who are under the control of the Director of the Country and Town Planning (DCTP), Chennai, showed a map which was prepared during 1994. When I asked for the latest map as the HTL and LTL might have undergone major changes due to the 2004 Indian Ocean tsunami, they started blinking. By that time the DC understood and pressed for the latest map. The officials replied that the latest map was available at their Tirunelveli office. When I asked them about the agency involved in preparing the latest map, they said it was Anna University, Chennai. With the available map, I was able to convince the DC that the area in question was well within CRZ-I. But if the latest map is referred, definitely the chances of the area falling under CRZ-I would be more. After listening to my discussion and inspection of the said area without any park like infrastructure, the DC made up his mind and came to a conclusion.

Subsequently the DCZMC meeting was held sometime during November, 2010, and a suitable order was passed by the District Coastal Zone Management Authority (DCZMA) granting environmental clearance under CRZ Notifications with certain conditions. Accordingly, only the Kanyakumari Town Panchayat could take up construction activities after demarcating the HTL as per the approved plan of the Anna University with proper stone pillars. It was to be ensured that the sea-facing area from the HTL would not be touched. No further permanent construction activity was to be carried out without prior approval of the DCZMA. The Kanyakumari Town Panchayat could give it to the Licensee only for maintenance and any other work would be undertaken at the cost of the Town Panchayat. No entrance fee was to be collected from any of the person visiting the park.

With so many conditions which are not in any way beneficial to the Licensee, the hotelier did not turn up for even maintaining the park. Thus, timely and prudent action taken while implementing the rules helped save the valuable marine environment and ecology.

Courtesy: Google

The Vivekananda Rock Memorial at Kanyakumari

http://www.downtoearth.org.in/blog/timely-and-prudent-action-60904

Tuesday 19 June 2018

8

A UNIQUE DRAGON

The population of the Southern Flying Lizard of the Western Ghats is fast dwindling and needs conservation.

It is a unique animal. A reptile that "flies". Well, almost. It actually glides from tree to tree in search of insects with the help of a wing-like membrane known as "patagium" attached to the sides of the body, supported by elongated ribs.

I am talking about the "Southern Flying Lizard", known to science as *Draco dussumieri*. "Draco" means "Dragon" in Greek and Latin. The specific name "dussumieri" honours Jean-Jacques Dussumier, a French voyager and merchant from Bordeaux. He collected zoological species from South Asia and regions around the Indian Ocean between 1816 and 1840.

Draco dussumieri was identified as early as 1837 and found mention in the catalogue of the world's reptiles. It is found in South India, specifically in the Western Ghats portion of the region-in states like Karnataka, Kerala, Tamil Nadu and Goa. In fact, it is the only species out of a total 42 grouped in the genus "Draco" that is found in Southern India. In recent years, it has also been reported in the Kanyakumari Wildlife Sanctuary of Tamil Nadu as well as some parts of the Eastern Ghats in Andhra Pradesh.

The lizards are arboreal (tree-dwelling). Besides hill forests, they are also found in nearby palm and areca nut plantations. Their home range consists of a few trees. Males are in the habit of maintaining small territories of two to three trees, which they patrol and chase away intruding males. Females move freely within the territory.

As mentioned earlier, the lizards move during the day from tree to tree in search of insects like ants and termites. They glide by extending the patagium- the flaps of skin attached to the side of the body that are supported by six elongated ribs. They can cover a distance of nearly 10 metres when they glide from one tree to another. The tail acts as rudder, giving direction to the flight. Gliding between the trees is also an escape mechanism.

Besides the fact that it "flies", the lizard is also unique for its colouring. Its colour is brown, with grey patches resembling the pattern of tree bark. Because of this advantage, they are very good at camouflage. As a result, it is very difficult to spot this elusive creature. In case of males, the underside of the wing-like membrane is blue in colour while the females have yellow colour. As the wing-like membranes on either side of the body shine with a brilliant hue, it is also known as "Butterfly Lizard".

These reptiles, being cold-blooded, warm up their body in the early morning sun in order to be active during the day time. Though they spend their entire lifetime on trees, females come down to the ground for laying eggs in the soil.

The lizards have a lot of predators. Arboreal snakes, birds like the Indian Golden Oriole and Black-capped Kingfisher and Lion-tailed macaques feed on them.

While humans do not feed on these little animals, their impact on them is seen in other ways. Due to habitat fragmentation caused by development because of hydroelectric projects, irrigation dams, railway lines, roads and plantations, there has been a decline in their population. As many of the forest fragments where they live are under the control of private owners, the protection level is highly negligible.

The Southern Flying Lizard is a unique animal. The aesthetic value of watching it in flight is very rewarding. Since this lizard species plays an important role in the food chain and its population is fast dwindling, due and urgent measures are to be taken to conserve them. It is very much necessary to promote the protection and restoration of this lizard species through education, economic incentives and inclusion of the Southern Flying Lizard under the Wildlife Protection Act, 1972.

Courtesy: Mr.K.Manikandan, Forester

Back and underpart of Flying Lizard with criss-cross veins

http://www.downtoearth.org.in/blog/a-unique-dragon-60993

Sunday 01 July 2018

9

TUMAKURU SETS A GREEN EXAMPLE

The hilly and forested district in southeast Karnataka is giving new lessons to the rest of India on responsible forestry.

Tumakuru district in south-eastern Karantaka, is something of an aberration. Located in the heart of the dry Deccan Plateau, it consists chiefly of elevated, forested land. Thus, in stark contrast to the landscape to its north, it is very green. Though inland (it is a one-and-a-half-hour drive from Bengaluru), it is known for the production of coconuts, because of which it is called as "Kalpataru Nadu" or "the Land of coconuts". And because Tumakuru has always been an exception rather than a norm, it is setting the trends in another field for the rest of India: responsible forestry.

Despite the fact that Tumakuru has a forest cover of 1,292.76 sq km i.e. 8.58 per cent of the total area of the district, with an annual rain fall of 575 mm, the Forest Department is making highly appreciable, untiring efforts in increasing the density of degraded forests and greening the highways and rural roads.

Funds for afforestation are sourced from various schemes like Karnataka Forest Development Fund (KFDF), Road Side Plantation (RSP), Karnataka State Highways Improvement Programme-Phase-II (KSHIP-II), Development of Degraded Programme (DDP), National Bamboo Mission

(NBM), Greening of Urban Area (GUA), Karnataka State Fund (KSF) and Mahatma Gandhi National Rural Employment Guarantee Act (MGNREGA).

But it is the innovative nursery and planting technology followed by Forest Department officials that is reaping dividends. For the greening of urban roads and highways, taller plants of 3 to 4 metres are raised in the nursery within a year by using bigger size containers, selected seeds of known origin and application of organic manure. Pits are dug in advance in the month of April along roads and highways, by using machinery. When the first rains arrive in May, planting is carried out. Before planting, the container plants are dipped in water so that the plant will have moisture for a considerable period of time after planting. The plants are provided with supporting sticks, usually 3 metres long and 15 cm in girth, to prevent them from swaying due to wind and tied with thorny phoenix leaf to protect them from cattle. With the available moisture, the plants get established. Irrigation facilities are provided for watering the plants during summer. More than 95 per cent of the plants get established well and those that don't are replaced in the subsequent year.

The degraded forest blocks are selected for planting under the KFDF scheme. Trenches measuring 4.2 metres in length, 0.50 metres in width and 0.50 metres in depth are dug with a space of 4 metres between each by using machinery. The container plants of 8"x12" are dipped in water before planting and 2 plants are planted in each trench with subsequent Digital Aerial Photogrammetry (DAP) application. Since the soil in the trench is loosened, it helps the plants' roots to penetrate deep and absorb soil moisture. With proper watch, the seedlings get well-established and the success rate becomes high. In spite of the careful planting and subsequent upkeep, there may be some casualties due to damage caused by porcupines and hares. Generally, this type of planting is carried out in state forest areas.

By using novel planting techniques, the officers of the Tumakuru Forest Division are doing a wonderful job of greening the district successfully and the same will increase the density of forests, leading to an improvement in the ecological health of the region. In stark contrast, plants of 2 to 3 metres height are planted in small-sized pits in other states. Moreover, the techniques employed to guard the trees are not effective, resulting in more number of casualties.

Who can mention Tumakuru without referring to Salumarada Thimmakka? The "Salumarada Thimmakka Tree Park" established in the Gubbi forest range, on the road which leads to Tumakuru, is in recognition of the great service rendered by the 106-year-old environmentalist, who planted over 8,000 trees in 65 years. The park plays a major role not only in greening the district but also in motivating students and other members of the public to love trees and in creating an interest in conserving forests.

The park was inaugurated by Karnataka's Minister of Forests, R Shankar on July 7, 2018. It has got 40 different species of plants with 2 pergolas (arched structures in a garden or park consisting of a framework covered with climbing or trailing plants), sitting benches, drinking water facilities, a 2.5 km long walk-way ringing the park from inside and a children's play area with the latest facilities. *Swietenia mahogani, Pongamia pinnata, Bambusa bambos, Dendrocalamus strictus, Madhuca indica, Syzygium cumini, Pterocarpus marsupium, Azadirachta indica, Dalbergia latifolia, Ficus glomerata, Ficus benghalensis, Ficus religiosa, Melia dubia, Cordia myxa, Thespesia populnea, Michelia champaca, Terminlia catappa, Lagerstroemia speciosa* and *Terminalia tomentosa* were planted on the park during the month of March and have established themselves well. The park can be visited without paying any entrance fee.

If such novel and scientific techniques as those followed by the Tumakuru Forest Division are emulated throughout the country, increasing India's forest cover and ecological health won't remain a distant dream.

Acacia auriculiformis tank foreshore plantation raised in Tumkur Forest Division

Photos by Author

Plantation raised under CAMPA in Bukkapatna Range of Tumkur Division

http://www.downtoearth.org.in/blog/tumakuru-sets-a-green-example-61113

Thursday 12 July 2018

10
THIS TOWN IN KARNATAKA HAS ITS VERY OWN CENTRAL PARK

The Gubbi Tree Park, inaugurated recently by the Tumakuru Forest Division in Karnataka, has shone a light on creating state-of-the-art green spaces meant for civic as well as forestry purposes.

The town of Gubbi is one of the 10 *taluk* headquarters of Tumakuru district in Karnataka. It is situated at a distance of about 20 km from Tumakuru city. Though a small town, with a population of approximately 20,000 at the moment, it has been expanding. The famous Hindu temple of Channabasaweshwara located in Gubbi attracts a large number of pilgrims throughout the year. Of late, many educational institutions have come up on the outskirts of town, increasing the basic needs of the public for a healthy life. Recently, the Tumakuru Forest Division decided to establish the Gubbi Tree Park with an idea of meeting basic civic requirements like walking tracks and playgrounds for children. The park was built on 30 out of 283 hectares of State Forest. Located on the Gubbi-Bidare road, the park is about 0.50 km from the Gubbi Railway Station and about 1.50 km from the town's Bus Station.

The site for the park was selected carefully, in the vicinity of the town, so that it could be easily approachable. The area, being a notified State Forest, already has tree growth and the same will have a positive impact on visitors till the saplings planted recently attain considerable size. By way of forming such parks in notified forests, possible encroachments by anti-social elements can be avoided. Utmost care is taken to select such areas which do not have any depredations by wild animals and feral monkeys. As the objective of the formation of this kind of Tree Park is to spread the concept of creation of woodlots (a parcel of a woodland or forest capable of small-scale production of forest products) all over the state, only one such site is selected in each district.

The entire area is protected by planting Reinforced Cement Concrete posts with a chain link fence all around. Walking paths with rain water drainage have been developed on either side of the park. Tree species which are endemic to the region have been given preference. Taller plants of a minimum height of 4 metres have been planted in bigger size pits with supporting poles in order to prevent swaying due to seasonal winds and avoid any possible damage to the younger saplings. Two low cost pergolas (a type of gazebo) using local materials have been erected for relaxation of the visitors.

Saplings of 40 tree species have been planted here with the application of organic manure. *Santalum album, Pterocarpus santalinus, Pterocarpus marsupium, Mesua ferrua, Butea monosperma, Annona squamosa, Cassia fistula, Thespesia populnea, Swietenia macrophylla, Swietenia mahogany, Saraca ashoka, Mangifera indica, Syzygium cumini, Hardwickia binata, Pongemia pinnata, Dalbergia latifolia, Azadirachta indica, Holeptelia integrifolia, Melia composita, Ficus benghalensis, Ficus religiosa, Ficus glomerata, Cordia myxa, Madhuca latifolia, Artocarpus heterophyllus, Tamarindus indica, Bambusa bambos, Dendrocalamaus strictus, Bauhinia purpurea, Michelia champaca, Tecoma stans, Terminalia catappa, Terminalia crenulata, Lagestroemia speciosa, Albizzia lebbek, Feronia elephantum, Aegle marmelos, Dalbergia sissoo, Simarouba glauca,* Singapore cherry, etc., have been planted in the tree park. Each plant will be provided with information boards with its scientific and vernacular names, the family it belongs to and its ecological, medicinal and commercial values in order to generate interest among the onlookers to grow more such trees for the betterment of the globe as a whole as trees play a major role in mitigating global warming.

According to Australian horticulturalist Nancy Beckham, 'Trees and plants silently carry out their daily routines year after year, stabilising the soil, recycling nutrients, cooling the air, modifying wind turbulence, intercepting the rain, absorbing the toxins, reducing fuel costs, neutralising sewage, increasing property values, enhancing social awareness, providing beauty, cutting noise, giving privacy, promoting tourism, encouraging recreation, reducing stress, and improving personal health as well as providing food, medicine and accommodation for other living things'.(www.uow.edu.au/~sharonb/STS300/ valuing /price/pricingarticles.html).

Drinking water and irrigation facilities have been provided in the Gubbi Tree Park. It is watered by a canal that has water during the rainy season. Sitting benches and dustbins have been provided all around the walkway and near the existing trees. Children's play areas have been formed with sufficient numbers of play equipment. There is a proposal to erect a few solar lights and low cost toilets for both, men and women. Entry to the Tree Park is free for the local public and students of schools and colleges.

There is a proposal to build an 'information centre' in the park. Once ready, it will harbour information about the great Native American naturalist Chief Seattle, the sacrifice of 365 Bishnois in Khejarli village in 1731 near Jodhpur in Rajasthan in order to protect *Prosopis spicigera* trees, the story of Gaura Devi of Garhwal's Rane village who protected trees from the axes of forest contractors during the Chipko Movement in 1974, the inspiring story of Jadav Payeng Molai of Assam, who single-handedly raised about 1,360 acres of forest and of course, the tale of Salumarada Thimmakka of Gubbi, who along with her husband have raised 385 banyan trees on either side of the road between Hulikal and Kudur to a stretch of 4 km.

In today's world, there is a shortage of green civic spaces in many cities globally. In this scenario, the establishment of the Gubbi Tree Park is of immense importance. As healthy trees create healthy communities, better business and higher property values, establishment of woodlots like this one may pave the way for peaceful and prosperous living. Developing tree parks in every district of India will not only benefit society but also lead to mitigation of ever-increasing global warming and climate change. In this respect, the Karnataka Forest Department is performing a wonderful job.

Visit to the Tree Park at Gubbi, Tumkur by school students

Photos by Author

The Author along with the Forest Officers of the Tumkur Forest Division in front of the Park

http://www.downtoearth.org.in/blog/this-town-in-karnataka-has-its-very-own-central-park-61224

Tuesday 24 July 2018

11
TO CREATE A WILD HAVEN

The Tumakuru Forest Division has submitted a proposal to the Karnataka government for declaring the Bukkapatna State Forest as a Wildlife Sanctuary. If adopted, it will be a boon for the biodiversity and ecological health of the region.

A "Protected Area" means a National Park, a Sanctuary, a Conservation Reserve or a Community Reserve notified under Sections 18, 35, 36-A and 36-C of the Wildlife Protection Act, 1972. When a notified State or Reserve Forest is declared as a Protected Area, it commands a better status. The protection is strengthened. The State and the Central Governments evince keen interest in developing such areas in order to improve the biodiversity therein which, in turn, provides all the ecosystem services such as carbon sequestration, release of oxygen, rain water harvesting, shelter for wildlife, soil conservation and supply of food and medicine.

The Tumakuru Forest Division, which operates in Karnataka's Tumakuru district, is taking a keen interest in bringing more State Forests under the "Protected Area" status, with the idea of improving the ecological health of the region. Its latest move in this regard is over the Bukkapatna State Forest.

Notified in the year 1900, Bukkapatna State Forest falls under the Bukkapatna Range of the Tumakuru Forest Division. It is considered to be one of the last remaining large, natural, open savannah woodland forests of

Karnataka. Major portions of the forest consist mostly of open grasslands, interspersed with natural vegetation. Certain parts of this forest have both, natural and man-made plantations of Hardwickia binata. The annual rainfall in this region is around 560 mm. The Range Forest Office of Bukkapatna had been built in 1913, during the British Era, considering the importance of the region. Renovated recently, without altering the original design, the office building has an aesthetic value to it. The Heritage Forest Rest House built at Bellara in 1911 near the state forest is also evidence of the real interest taken by British forest officers in protecting the forest wealth.

But the real wealth of Bukkapatna is its animal wealth. About 18 mammal species have been documented through camera traps in this forest. Some of the "Vulnerable" and "Near Threatened" faunal species as per the International Union for Conservation of Nature (IUCN) which are found here include the Common Leopard, Sloth Bear, Chinkara or Indian Gazelle, Chowsingha or Four-horned Antelope, Blackbuck and Striped Hyena respectively.

Though Chinkara were recorded in Bagalkot district earlier, this is the first time they have been recorded in Tumakuru district. About 200 Chinkara are estimated to reside in the area. In addition to Chinkara, two other antelope species including Chowsingha and Blackbuck have also been recorded here. The vast extent of open grasslands, with sparse tree growth as preferred by these antelopes may be a reason for the presence of them in this forest. The Striped Hyena, which prefers this kind of forest, has also been recorded here.

Other mammal species that are also captured by camera traps include the Jackal, Jungle Cat, Rusty-spotted Cat, Wild Boar, Common Mangoose, Ruddy Mangoose, Black-naped Hare, Bonnet Macaque, Southern plains Grey Langur, Indian Porcupine, Small Indian Civet and Common Palm Civet or Toddy Cat.

Based on a study made by the Tumakuru Forest Division, a proposal has been submitted to the Government of Karnataka for declaring the Bukkapatna State Forest with an extent of 13,188.70 ha as a Wildlife Sanctuary as such an assemblage of dry species in one area is rare. Above all, all the four species, including Chinkara, Four-honed Antelope, Blackbuck and Hyena have been protected under Schedule-I of the Wildlife Protection Act, 1972. If the proposal is accepted and suitable order passed, it will be a boon not only for the denizens of the dry forest, but also for the whole of Karnataka, as it will improve the biodiversity and the ecological health of the region.

Chinkara or Indian Gazelle in Bukkapatna forest

Courtesy: Bukkapatna Forest Range

Photo by Author

Heritage Forest Rest House built in 1911 at Bellara

http://www.downtoearth.org.in/blog/to-create-a-wild-haven-61391. Tuesday 14 August 2018

12
NEELAKURINJI: BLUE BLOOM IN A BLUE MOON

2018 is the year when the shrub is flowering again in the hills of South India. Here is its story.

Kurinji or *Neelakurinji*, scientifically known as *Strobilanthus kunthianus*, is a shrub that grows in the *shola* forests of the Western Ghats in South India. The plant is named after the famous Kunthi River which flows through Kerala's Silent Valley National Park, where the plant occurs abundantly. The Kurinji plant belongs to the genus *Strobilanthus*, family *Acanthaceae* and was identified in the 19th century. The genus has about 250 species. Out of that, around 46 species are found in India. Kurinji grows to a height of 30 to 60 cm and is found at an altitude of 1,300-2,400 metres.

Most of the *Strobilanthus* species have an unusual flowering behaviour varying from an annual to 16-year blooming cycles. Characteristics include gregarious flowering, mass seeding and synchronised monocarpy (the characteristic character of certain plants which flower once in their lifetime and die after fruiting).

Honey bees act as pollinators of *Neelakurinji*. The nectar collected by honey bees from these flowers is found to be very tasty, nutritious and has medicinal values.

Some Kurinji plants bloom once in every seven years and then die. Their seeds sprout subsequently and continue the cycle of life before they die eventually. *Strobilanthus kunthianus* and other species, that are long interval bloomers, are known as "Plietesials" scientifically. *Strobilanthus kunthianus* blossoms only once in 12 years. The blooming of this plant has been documented in 1838, 1850, 1862, 1874, 1886, 1898, 1910, 1922, 1934, 1946, 1958, 1970, 1982, 1994, 2006 and 2018.

Kurinji has long featured in the culture of South India, especially the modern-day states of Kerala and Tamil Nadu. In the ancient Sangam literature of Tamilakam or Tamil Country, land was classified into five types. They are *Kurinji* (mountainous), *Mullai* (forested), *Marutham* (agricultural), *Neithal* (coastal) and *Paalai* (desert). Tamil scholars opine that this classification was based on the most characteristic plants of these ecosystems: *Strobilanthus kunthianus* (Kurinji), *Jasminum auriculatum* (Mullai), *Nymphaea nouchali* (Neithal) and *Wrightia tinctoria* (Paalai). The mountainous landscape, referred to as Kurinji, abounded with Kurinji flowers. The Paliyar tribal community that lives in the montane rain forests of the South Western Ghats uses the flowering periodicity of this plant to calculate their age.

Kurinji used to once grow abundantly in the Nilgiri Hills (part of the Western Ghats) in Tamil Nadu. The brilliant blue colour of Kurinji has given the hills the name "Nilgiri", literally meaning "Blue Mountains". But presently, plantations and buildings have occupied the hills. In Kerala, the Anamalai Hills of Idukki district, the Agali Hills of Palakkad district and the Eravikulam National Park of Munnar (all in the Western Ghats) also have this plant. In addition to the Western Ghats, Kurinji is also found in the Shevaroy Hills of the Eastern Ghats in Tamil Nadu as well as the Bellary district of Karnataka.

The year 2006 was when the Neelakurinji last bloomed in Kerala and Tamil Nadu, after a span of 12 years. The year was declared as the "Year of Kurinji" and a commemorative stamp was released in Kerala. There is a sanctuary in Kottakamboor and Vattavada villages of Idukki district specially meant for conserving Kurinji called "Kurinjimala Sanctuary".

Twelve years after 2006, the Kurinji is in full bloom again in South India from July this year. It will continue up to the month of November. The best place to see the plant and its blossoms is the hill station of Kodaikanal in Tamil Nadu. The name "Kodaikanal" means "Gift of the Forest" in Tamil. It is referred to as the "Princess of hill stations" and is frequented by tourists, especially during the summer. Kodaikanal is situated on the crown of the Palani Hills, at an altitude of 2,133 metres (6,998 feet) and is surrounded by dense forests. The Paliyar tribes are considered to be the earliest residents of Kodaikanal. References about this enchanting hill station are found in Sangam literature. The famous Kurinji Andavar Temple in Kodaikanal, is named after Kurinji, that carpets the region every 12 years.

Since awareness about Kurinji is not widespread other than among nature lovers, research students and forest officials, the Dindigul district administration (where Kodaikanal lies) has planned to celebrate a grand festival on Kurinji this year in order to make the public aware of this wonderful flowering species. Hoardings will be erected with details of the plant species in the regions where the gregarious flowering is seen. The festival on Kurinji has been planned for this month.

Kurinji is a species that deserves our attention. We have only 10 per cent of *Strobilanthus* in Tamil Nadu and Kerala. The habitats of this plant species, such as the *shola* forests and the grasslands have been converted into tea and coffee plantations. Indiscriminate planting of exotic species like Pinus, Wattle and Eucalyptus on a large scale also have encroached upon the original habitats of this rare plant species. Increase in tourism, encroachment, water depletion and deposition of plastic waste have further degraded the ecosystem.

As Neelakurinji or *Strobilanthus kunthianus* occurs in grassland and *shola* forests, at an altitude of 1,300 to 2,400 metres, it is very essential to maintain and improve the ecosystem without any further degradation and depletion.

Neelakurinji in full bloom in Kodaikanal Forest Division

Courtesy: Kodaikanal Forest Range

Blue Bloom in a Blue Moon

http://www.downtoearth.org.in/blog/neelakurinji-blue-bloom-in-a-blue.moon-61295 Tuesday 02 August 2018

13
A UNIQUE TREE PARK FOR CHENNAI

With 300 species of trees, the 'Forest Genetic Resources Tree Park' will serve as a unique gene pool for Tamil Nadu's threatened flora.

The Tamil Nadu government recently opened a unique "tree park" in Chennai in an effort to save the state's unique flora.

On August 16, 2018, Minister of Forests, Dindigul C Srinivasan inaugurated the "Forest Genetic Resources Tree Park" at Kolapakkam, adjoining the Arignar Anna Zoological Park located at Vandalur, Chennai.

The taxonomically-designed tree park is the first of its kind and would be a haven for researchers, naturalists and botanists. The entry will be free for children and Rs.5/ for adults and the fee collected would be used for maintaining the park in the future.

The park was inaugurated as part of the centenary celebrations of the research wing of the Forest Department. It has been formed as part of a move by the Tamil Nadu government to save about 230 types of plants as announced a few years ago by the then-Chief Minister J Jayalalithaa in the state Assembly. She made a *suo motu* statement in the House, proposing the park in order to ensure protection of forest genetic resources. Her initiative has now been given shape.

The park is spread over 20 acres at Kolapakkam. It has a collection of 300 tree species from both, the Western and the Eastern Ghats. Each tree species has been provided with a hoarding containing the scientific and vernacular names and its commercial and medicinal uses. Seeds of some rare varieties of trees have been procured from the Kerala Forest Research Institute (KFRI), Peechi. Many tree species have been supplied by the Andhra Pradesh and Karnataka Forest Departments. The park is, thus, a kind of gene pool garden formed under *ex-situ* conservation (an assemblage of different floral species in a particular place from different regions) method.

The park is located between the Vandalur Reserve Forest (RF) and the Nedunkundram Lake. Rainwater drains from the Vandalur RF into the lake through the park. Three check dams have been built across the naturally-formed canal that passes through the park.

The water stored in the check dams will be made use of for watering the tree species planted here besides improving the ground water level in the nearby areas. The interpretation centre constructed here may have displays of various rare, endangered and threatened tree species besides depicting the values of trees and the ecosystem services that they provide for the benefit of all living organisms on earth. The student community, especially natural science students may make use of this to enrich their knowledge on plants and other valuable biodiversity.

There is also a Rock Garden, Butterfly Garden, Children's Garden, Arboretum, Bambusetum, Palmatum, Flower walk Garden, Medicinal Garden and Water Cascade, that will serve as additional attractions for visitors.

Courtesy: Google

Forest Genetics Resources Tree Park developed at Chennai

http://www.downtoearth.org.in/blog/a-unique-tree-park-for-chennai-61488

Friday 31 August 2018

14
A HISTORIC JUDGEMENT BY THE SUPREME COURT ON ELEPHANT CORRIDORS

By ordering the shutdown of illegal resorts built on the Sigur Plateau corridor in the Nilgiris, the apex court has set a precedent for more effective pachyderm-protection laws in the future.

Last month, the Supreme Court (SC) did something extraordinary. On August 9, in response to a Public Interest Litigation, it directed the Tamil Nadu government to seal or close down 39 hotels and resorts constructed on an "elephant corridor" in the Nilgiri Hills in violation of law, within the next 48 hours.

Justice Lokur, along with Justices S Abdul Nazeer and Deepak Gupta also said elephants were the country's "national heritage" and expressed displeasure about the encroachments.

It was a historic judgement by the SC on India's elephants. While they play a key role as a "Keystone Species" in the forest ecosystem and are termed as the "National Heritage Animal of India" by the Union Ministry of Environment, Forest and Climate Change, the plight of the Indian Elephant is known to all. As more forest areas face fragmentation due to unscrupulous encroachment and other developmental activities, the pachyderms are forced to migrate and find themselves in human-dominated landscapes, leading to human-elephant conflict. Cattle grazing, coffee and tea plantations, private tourist resorts, electric fences, expansion of agricultural fields, vehicular traffic especially during night hours and human settlements along elephant corridors pose a greater threat for the free movement of elephant herds that are in the habit of migrating across 350-500 sq km annually.

The August 9 judgement specifically dealt with elephant corridors. What are they? Elephant corridors are narrow strips of land that connect two large habitats. In a 2017 study, Delhi-based non-profit, Wildlife Trust of India (WTI), in collaboration with "Project Elephant" and UK-based non-profit "Elephant Family", has identified 101 elephant corridors in India. Out of these 101 corridors, 28 have been identified in South India, 25 in Central India, 23 in North-Eastern India, 14 in northern West Bengal and 11 in North-Western India. While 70 per cent of the 101 corridors are regularly used, 25 per cent are used occasionally and 6 per cent rarely. In general, 93 per cent of the elephant corridors in South India, 86 per cent in northern West Bengal and 66 per cent in north-eastern India are regularly used.

But here is the worrying part. The study reports that about 74 per cent corridors have a width of one kilometre or less today, when compared with 45.5 per cent in 2005, and 22 per cent corridors have a width of one to three kilometres now when compared with 41 per cent in 2005, showing how the corridors have become narrower. While about 400-500 human casualties are caused by elephants and 100 elephants are killed in retaliation annually, it is really high time that effective action is needed in saving the elephant corridors that are traditionally used by the elephants.

The SC judgement dealt especially with the Nilgiri elephant corridor of Tamil Nadu. That corridor is the Sigur Plateau, with a width of 1.5 kilometres and length of 22 kilometres. It connects the Western and the Eastern Ghats and sustains elephant and tiger populations and their genetic diversity. The Plateau has the Nilgiri Hills on its south-western side and the Moyar River Valley on its north-eastern side. The South-West monsoon is active between

June and September in the Western Ghats and the North-East monsoon is active between October and January in the Eastern Ghats.

Depending on the rainy seasons, the elephants migrate in search of food and water and during the course of their migration, they have to cross the Sigur Plateau. This elephant corridor acts as an important link connecting the protected areas of Tamil Nadu and Kerala. This migratory path is considered to be very crucial connecting several contiguous protected areas forming the Nilgiri Biosphere Reserve, the largest protected forest area in India. This reserve supports over 6,300 elephants. The Nilgiri Biosphere reserve, which includes Sigur Plateau and the Nilgiri Hills, is part of the UNESCO World Network of Biosphere Reserves.

The private resorts mentioned in the SC judgement, pose a serious threat of fragmentation and destruction of habitats, loss of connectivity between habitats due to construction of new buildings and erection of barbed and electric fences. Besides the local tour vehicle operators, the resort owners also indulge in operating night safaris, violating the provisions of the Indian Wildlife Protection Act, 1972. The jeeps and trucks which are engaged in taking the tourists for watching the wildlife have also increased manifold in recent years.

There are plenty of laws to protect the Sigur corridor as well as others in Tamil Nadu. As most of the private lands are covered under the Tamil Nadu Preservation of Private Forest (TNPPF) Act, sale of land and any development without proper permission are considered violations. If any such violations are noticed, the District Collector is empowered to cancel the sale deed in case of any sale proceedings. Under this scenario, the new constructions in the name of resorts also attract the provisions of the TNPPF Act. The Hill Area conservation Authority (HACA) is also being implemented in this region. As many resorts and other developmental activities have been carried out without proper permission of the Director of Town and Country Planning, they are liable for penal action.

However, these laws, till now, had not been able to counter the menace of encroachments in the corridor. Reason: Lack of implementation and loopholes. For instance, in certain cases, the local Panchayat or the Executive Officer was in the habit of issuing permissions. But legally speaking, this kind of permission is unauthorised.

This is where the SC's judgement sets a precedent. After the apex court's order, the decade-long battle over illegal constructions, encroachments and other disturbances in the Sigur valley elephant corridor will come to an end allowing a free walk for the jumbos in their regular path of migration. It also sets the pace for better and more effective laws for the protection of elephant corridors across India.

A herd of wild elephants

Courtesy: Google

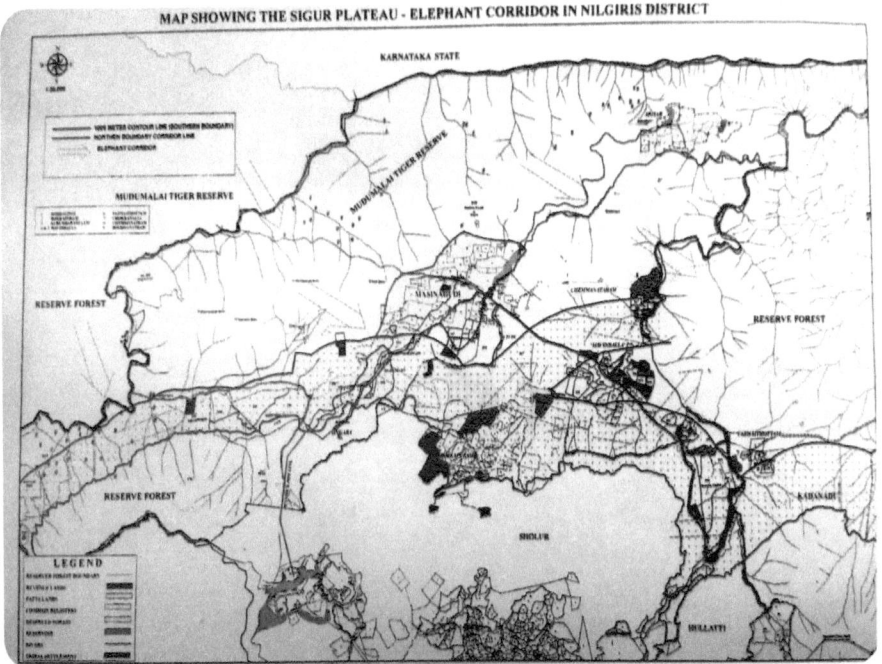

Courtesy: Nilgiris North Forest Division

Map showing the Sigur plateau-Elephant Corridor in Nilgiris District

https:/www.downtoearth.org.in/blog/a-historic-judgement-by-the-supreme-court-on-elephant-corridors-61585

Monday 10 September 2018

15
E-WASTE: HOW SHOULD INDIA DEFUSE A TICKING TIME BOMB?

Safe disposal practices, consumer awareness and cooperation between organised and unorganised sectors are needed to solve the issue.

India's annual electronic waste (e-waste) generation was 1.8 million MTs in 2016 and is expected to reach 5.2 million MTs by 2020. Mumbai stands first among the top ten Indian cities generating e-waste, followed by Delhi, Bengaluru, Chennai, Kolkata, Ahmedabad, Hyderabad, Pune, Surat and Nagpur. The government, public and private sectors act as the primary source of e-waste, accounting for 70 per cent. Individual households contribute only 15 per cent. The balance 15 per cent is produced by manufacturers.

When this is the state of affairs vis-a-vis e-waste in India, it is high time the Union government rises to the occasion to prevent the increasing environmental damage and health hazards caused by such waste.

But first, let us understand what exactly e-waste is. Electronic and electrical equipment that have become unfit for their originally intended use or which have crossed the expiry date are called "e-waste". Computers, servers, mainframes, monitors, CDs, printers, scanners, copiers, calculators, fax machines, battery cells, cellular phones, transceivers, TVs, iPods, medical apparatus, washing machines, refrigerators and air-conditioners are examples of e-waste.

If e-waste is processed scientifically, valuable metals such as copper, silver, gold and platinum could be recovered from it. However, substances like liquid crystal, lithium, mercury, nickel, polychlorinated biphenyls (PCBs), selenium, arsenic, barium, brominated flame proofing agent, cadmium, chrome, cobalt, copper and lead, which are an inherent part of electronic equipment, are toxic and carcinogenic. If e-waste is dismantled and processed in a crude manner, its toxic constituents can wreak havoc on the human body.

The cathode ray tubes (CRTs) present in computer monitors, with heavy metals like lead, barium and cadmium, may be harmful during the improper processing and cause an adverse impact on the human nervous and respiratory systems. In the same way, lead and cadmium present in the printed circuit boards, beryllium of the motherboards, mercury in switches and flat-screen monitors, cadmium in the computer batteries, polyvinyl chloride (PVC) in the cable insulation and bromine in plastic housing may cause damage to the human body parts such as nervous system, kidney and liver, lungs and skin, heart, lever and muscles, brain and skin, kidney and liver, immune system and endocrine system respectively.

In India, 90 per cent of the recycling and disposal of e-waste is done by the informal/unorganised sector. Unskilled workers not only work without any protection measure or safeguards but also live in slums close to the untreated e-waste dumps and landfills. For instance, they don't wear any glove or mask when they use nitric acid for removal of gold and platinum from the circuit boards. Usually, children are engaged in dismantling the circuit boards.

The private and public sectors prefer selling their e-waste to informal dismantlers as they get more price since the expenditure is less when the recycling is done through the unorganised sector.

Solutions

The generation of e-waste in large quantities in recent years poses a serious threat to the environment and human health. What can be done to tackle the problem?

First and foremost, inventorisation of the e-waste produced annually should be done by engaging an established government agency. Only then suitable steps can be planned for recycling and disposal of e-waste in an organised manner.

When valuable metals are extracted from e-waste, it can meet out our domestic requirement. In that way, our dependence on other countries for the import of such metals can be dispensed with. As e-waste contains toxic components, it is very much essential to segregate and dispose them scientifically and carefully in order to manage the environment sustainably.

Awareness is key both for the stakeholders and consumers. Consumers also should understand certain intricacies. When they buy any electrical/electronic product, it may be somewhat expensive. However, after certain years of use, its lifetime may expire. At that time, it is better to dispose of them properly by selling such products to the organised sector. During that time, the product becomes scrap, and only the metal present may have some value. So, naturally it may fetch only a small price. But consumers are hesitant to part with the outdated product thinking of the original price they paid as well as some sentimentality attached to the product.

Awareness programmes need to be conducted to make the public understand the real health problems caused by untreated e-waste and the value of the components, and guide them to dispose the e-waste properly, through an authorised sector only. With the Indian government promoting the "Swachh Bharat" programme the country must be kept free of any untreated e-waste also.

Considering the adverse impacts caused by untreated e-waste on land, water and air; the government should encourage the new entrepreneurs by providing the necessary financial support and technological guidance. Establishment of start-ups connected with e-waste recycling and disposal should be encouraged by giving special concessions.

The unorganised sector has a well-established collection network. But it is capital-intensive in case of organised sector. Therefore, if both the sectors coordinate and work in a harmonious manner, the materials collected by the unorganised sector may be handed over to the organised sector to be processed in an environment-friendly way. In this kind of scenario, the government can play a crucial role between the two sectors for successful processing of the e-waste.

It is high time that the government takes a proactive initiative to recycle and dispose of e-waste safely to protect the environment and ensure the well-being of the general public and other living organisms.

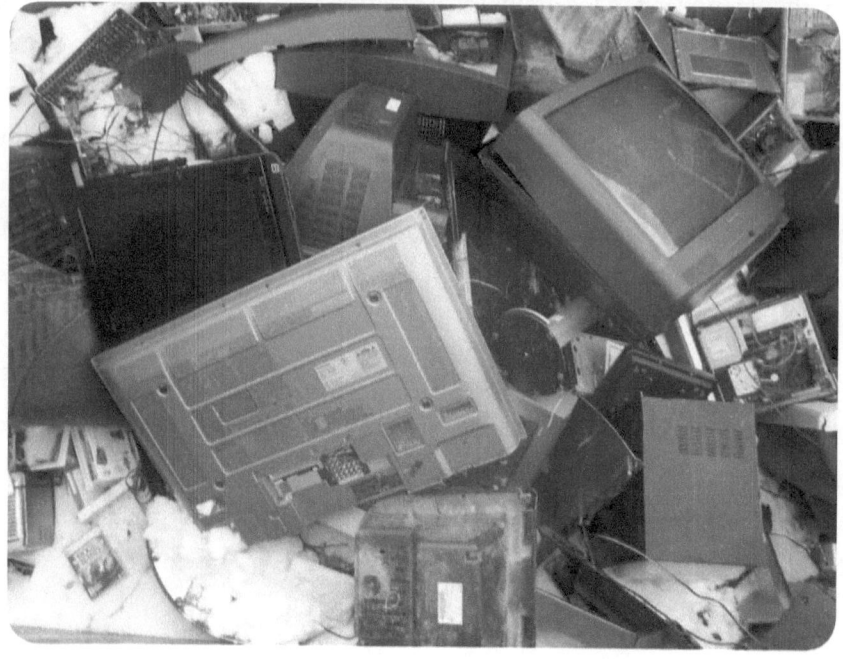

Courtesy: Google

Electronic waste generated

https://www.downtoearth.org.in/blog/e-waste-how-should-india-defuse-a-ticking-time-bomb-61617

Thursday 13 September 2018

16

HOW THESE TWO REMOTE DISTRICTS IN ANDHRA ARE TACKLING WILD ELEPHANT INTRUSIONS?

The forest department in Srikakulam and Vizianagaram, which border Odisha, are doing their best to manage pachyderms that regularly stray from Odisha. But they need local support.

On March 28, 2018, people in Srikakulam, Andhra Pradesh, were startled to hear that eight wild elephants had come to the Dusi Railway Station, about 10 km away from Srikakulam town. Forest officials immediately jumped into action. Using "Kumkis" (domesticated and trained elephants) brought from the Chittoor Forest Division, they were able to chase the jumbos back to Odisha, from where they had come. However, almost four months later, on July 26, the same herd returned. This time, tragedy struck when one young elephant got electrocuted after coming into contact with a transformer accidentally.

The above lines give you an idea of the tension that prevails in this area due to interaction between man and elephant. But this was not always the case. Elephants started moving from Odisha to the two border districts of Andhra (Srikakulam and Vizianagaram) only from 2007 onwards. They usually move from State Forests in Odisha to the Vizianagaram Forest Division and then migrate towards Srikakulam Forest Division.

The reason for the migration of elephants from Odisha to Andhra Pradesh is said to be large-scale quarrying permitted in the revenue hills contiguous to the Reserve Forest areas in Odisha districts bordering Srikakulam and Vizianagaram.

In general, permission for quarrying even in revenue and private lands is not granted if such areas are within 10 km of the boundary of any Protected Area like a Wildlife Sanctuary or a National Park. This rule should be made applicable to revenue hills and hillocks which are located close to the Reserve Forests where the elephants live. If this system is followed, chances of migration of the pachyderms due to human disturbance may be very much reduced.

However, that being not the case for now, sincere efforts are being made by the Forest, Revenue and Police officials to mitigate human-elephant conflict in the area. There are usually two kinds of mitigation: prevention and response.

Prevention refers to the measures that are taken to not attract elephants to the area or to see to it that even if they come, no human being crosses paths with them.

In many parts of the agricultural lands that are close to the boundary of Reserve Forests in Srikakulam and Vizianagaram, crops like sugarcane, banana, pineapple and maize are raised, which in turn, attract the pachyderms. People, especially the farming community who are cultivating close to the Reserve Forests inhabited by elephants, are persuaded not to raise such crops. Instead, they can raise crops like garlic, turmeric, mulberry, castor, chilli, cotton, ginger, onion, tobacco and tea.

Around a few Reserve Forests, where elephant movements are noticed, Elephant Proof Trenches are dug out to a depth of 3 metres, with a top width of 3 metres and bottom width of 2 metres, with the idea of preventing the pachyderms from straying into the nearby agricultural lands and human habitations.

In order to see that humans do not cross paths with elephants if they do make it to human settlements, forest officials monitor elephant movements with the help of tribal elephant trackers and alert adjoining villagers. Flex boards have been erected on the roadsides and at junctions where elephant movements are reported. Pamphlets are distributed widely in villages. Publicity is also done through megaphones in vulnerable villages. Dos and don'ts are explained to villagers. Local tribes are being warned to be cautious.

Under response, whenever elephants come close to human habitations, various methods are used to drive them away. The elephants are frightened by beating drums, firing crackers and burning tyres. In addition, there are eco-friendly measures too. For instance, gunny bags smeared with chilli powder mixed with waste crude oil are tied with a rope. The rope is tied to strong poles, which are erected across the pathways by which the elephants would cross from the RF into the nearby fields. Besides gunny bags, beehives can also be suspended from the rope. When the jumbos come into contact with the rope, the beehives are shaken and as a result, the bees come out and scare the elephants. The beehive method is followed successfully in Kenya.

Another aspect of the administration's mitigation techniques involves compensation. In case of any human death caused by wild elephants, the state forest minister visits the village immediately and expresses his condolence to the bereaved family, besides paying due compensation to the legal heir of the diseased. During 2007, the compensation was Rs.1 lakh in case of human casualty and the same amount was awarded till 2014. From 2015 onwards, the compensation was increased to Rs 5 lakh. When agricultural crops like paddy, sugarcane, red gram and banana, and horticultural crops like cashew and mango are damaged by elephants, the value of the damage is assessed through the concerned agriculture and horticulture departments and due compensation is paid to the affected parties.

The arduous efforts taken by the Forest, Revenue and Police Departments in these two districts are noteworthy. However, these alone will not be able to mitigate human-elephant conflict unless there is good support and cooperation from the villagers concerned.

Courtesy: Mr.Subbarao, Srikakulam

Wild Elephants being chased away with the help of "Kumkis" in Srikakulam

http://www.downtoearth.org.in/blog/how-these-two-remote-districts-in-andhra-are-tackling-wild-elephant-intrusions-61833

Tuesday 09 October 2018

17

GREEN AND GOLDEN SRIKAKULAM

Andhra Pradesh's remote, coastal district has turned green and is employing its inhabitants through a number of schemes started by government departments.

Srikakulam district in Andhra Pradesh lies on the coast of the Bay of Bengal. Scheduled Tribes form 6.15 per cent of the district population as per the 2011 Census of India. In the past few years, Srikakulam has set in place pioneering measures to make itself greener and provide viable opportunities to its economically weaker sections for their upliftment.

The district has done this through a 20-Point Programme that it carried out during 2016-17. The programme was implemented by various arms of the district and state administration namely, the Forest Department, District Water Management Agency (DWMA) and Integrated Tribal Development Agency (ITDA) that have implemented afforestation measures under a number of Central schemes.

For instance, the Territorial Forest Division (TFD) has raised teak, miscellaneous and medicinal plantations on 100 hectares of Reserve Forest (RF) under the Compensatory Afforestation Management and Planning Authority (CAMPA) and National Medicinal Plants Board (NMPB) schemes in order to improve biodiversity. The TFD has also created casuarina and palmyrah plantations along 80 hectares of the district's seashore under the National Afforestation Project (NAP) and the Andhra Pradesh Disaster

Recovery project (APDRP) schemes in order to protect inland areas from the fury of natural disasters like cyclones and tsunamis.

The Social Forestry Division (SFD) has developed avenue plantations under the Special Development Package (SDP) and the Mahatma Gandhi National Rural Employment Guarantee Scheme (MGNREGS) along 90 km of urban roads throughout the district and distributed 25 lakh casuarina, 13 lakh eucalyptus and 3 lakh other miscellaneous plants under MGNREGS to farmers and government institutions. The trees that have been planted include shade-bearing and flowering plant species like coconut, mahogany, neem, jacaranda, gulmohar and others.

The DWMA has supplied about 6 lakh horticulture plants to more than 87,000 farmers in over 38 *mandals* of the district for planting. These include species like cashew, mango, lemon, coconut, curry leaf, guava, jackfruit, custard apple, sapota and teak. Cashew, coconut and mango have been supplied to the farmers in order to augment their revenue. At the same time, other species like lemon, curry leaf, guava, jackfruit, custard apple and sapota have been supplied to individual farmers for raising homestead plantations.

The DWMA has also created avenue plantations along 570 km of rural roads with shade-bearing and flowering tree species such as Indian Beech, Indian Tulip, Indian Cherry, African locust bean, Indian Elm and Gulmohar.

Meanwhile, the Integrated Tribal Development Agency (ITDA) is concentrating on the development of Srikakulam's tribals through tree planting, by supplying horticulture plant species like mango and cashew, with 100 per cent subsidy for planting in 20 Micro Water Sheds (MWS) and paying Rs0.50/plant/daily for their maintenance for 3 years. Though this is dry land horticulture, purely depending on rain, the tribal farmers cultivate traditional crops like cotton, pigeon pea, etc., as intercrop in the plantations in order to augment their revenue.

Thus, by creating greenery all over the district through tree planting, not only has the tree cover increased, but people belonging to economically weaker sections also are being provided with an opportunity to augment their income. Horticultural plants such as mango, cashew and coconut may start yielding after 3-4 years, while other quick growing species like curry leaf, custard apple, lemon, guava and sapota fruit within 2-3 years, enhancing their revenue, leading to economic empowerment. Especially in case of tribals, the subsidy given initially and the financial aid provided subsequently for the maintenance of horticulture plants will definitely pave the way for leading a comfortable life.

Though the district has got only 12 per cent of forest cover, due to the massive tree planting activities undertaken by different state departments, the tree cover has increased considerably, enabling the area to attain ecological sustainability. The rainfall, ground water table, and the greenery found all over the district bear testimony to this. The district enjoys an annual rainfall of 1075 mm. The average ground water level has risen from 7.03 metres to 3.96 metres in recent years. Vast stretches of paddy fields and hills covered with dense green vegetation are visible in all the directions, except the North, where the sea coast is located. The dedicated activities of various agencies in making the district green through tree planting involving the economically weaker sections of the society have thus helped in not only just ecologically sustaining the district but also provided for the economic development of the marginalised sections of the society.

Photo by Author

A Cashew plantation raised under ITDA in Bhamini Mandal

http://www.downtoearth.org.in/blog/green-and-golden-srikakulam-61920

Tuesday 23 October 2018

18

TAMIL NADU'S INNOVATIVE METHODS OF SEED PLANTATION

The state has been trying out a number of novel and unique ways to plant seeds of various tree species and ensure that they survive and make waste and barren areas greener.

While the percentage of forest cover all over the world is 30.6, India's forest and tree cover has been assessed to be 24.39 per cent, consisting of 6,778 sq km of forest cover and 1,243 sq.km of tree cover. An analysis of Indian forest cover reveals that 2.99 per cent is "very dense forest", 9.18 per cent is "moderately dense forest" and the remainder is "open and scrub forest".

However, both, experts of forestry as well as India's National Forest Policy state that a country should have at least one third or 33 per cent of its land area under forests for ecological sustainability.

India has made efforts to achieve the target of 33 per cent. But an explosion of its population as well as human activities such as urbanisation, industrial and agricultural expansion, dam building and road laying mean that achieving it is going to be very difficult, if not impossible. Consequently, in many areas, increasing the tree cover in addition to protecting forest cover has now taken priority. Many schemes are afoot to increase tree cover by planting trees in waste and barren lands with the consent of the owners concerned.

In this context, Tamil Nadu has experimented with a number of ideas. In order to encourage tree cultivation outside forests, "Tree cultivation in Private Lands" or TCPL was implemented in the state from 2007-08 to 2011-12, which is the first effort of its kind in the country. An amount of Rs.56.96 crore was sanctioned under state funds from 2007-08 to 2011-12. It was implemented by carrying out block planting and inter-crop planting during 2007-08, by carrying out block planting during 2008-09 and 2010-11 and free supply of profitable timber species for planting in the bunds. During 2011-12, block planting was carried out by planting profitable timber species like teak and Casuarina species in farmers' lands.

The state has especially introduced new ways to sow seeds in order to ensure maximum chances of survival. In recent years, the technique of "seed ball" sowing has picked up in Tamil Nadu. Although it is an ancient technique that has roots in Antiquity—seed ball sowing was in use in Ancient Egypt, on farms, after the annual flooding of the Nile—it has become familiar after it was rediscovered by the Japanese natural farming pioneer, Masanobu Fukuoka.

Seed balling was first carried out in Tamil Nadu in 1987. At the time, aerial seeding of pellets was carried out by using pellets. The seeds were inserted inside earthen balls, the size of a plum. It was considered to be a modern technique of creating green cover and it yielded good results. The seed ball sowing method is also considered to be more cost-effective when compared with other methods.

In August this year, the technique was used again, in the state's Tiruchirapalli Forest Division. Seed balls, the size of an "Amla" or Indian gooseberry, were prepared from a mixture of red soil, cow dung, cow urine and compost. Seeds of tree species such as Indian Beach, East Indian Walnut, Margosa, Indian Cherry, Peepul, Banyan, Custard Apple and Tamarind were inserted in the earthen balls. The semi-solid balls were shade-dried initially and then allowed to dry in the hot sun for some time. They were then sown in the Tiruchi Forest Range and on the Perumal Malai hills, a revenue hill located near Thuraiyur.

In addition to Tamil Nadu, seed balling is being followed successfully in other southern states like Telangana, Karnataka and Andhra Pradesh. The Indian government should also promote this method.

Other seed sowing methods used or being used in Tamil Nadu include the transplantation method, the direct sowing method and the pre-treated seeds method.

In the transplantation method, 6 months to one-year-old seedlings are raised in the nursery and transplanted in pits of varying sizes of 0.45 cm to 1.00 m, depending on the size of the seedlings. In case of Teak, the seedlings are raised in the nursery and stumps are prepared by removing the adventitious roots and clipping at the collar level. Then, these stumps are planted during the monsoon in crowbar holes. This is called the stump planting method. In Tamil Nadu, this system of planting of Teak was followed in the past. Presently, this method of planting is followed in Andhra Pradesh.

Direct sowing of seeds is also done. But the method depends on the tree species. In case of Wattle, a type of Acacia, the direct sowing of pre-treated seeds was followed in the past in the Nilgiris district. As of now, this system has been abandoned as wattle is considered to be an exotic species in India. Pre-treated seeds were sown in holes made by sticks also in the past. A notable practitioner of this method was the famous England-born Indian Forest Service Officer Hugo Francis Andrew Wood. He planted teak seeds in the holes made by his silver-tipped walking stick at Topslip, presently part of the Anamalai Tiger Reserve in Tamil Nadu from 1918 to 1926. He lies buried among the Teak trees that he planted, with a Latin inscription on his tomb that reads "Si monumentum requires circumspice", meaning "If you want to see me, please look around".

Courtesy: Thuraiyur Forest Range

Seed balls kept ready for sowing

http://www.downtoearth.org.in/blog/tamilnadu's-innovative-methods-of-seed-plantation-62145

Thursday 15 November 2018

19

HOW ALIEN INVASIVE PLANT SPECIES THREATEN WESTERN GHATS?

The ecological equilibrium of an ecosystem can be maintained only by balancing the floral and faunal population. But often ecosystems like shola forests, evergreen forests, grasslands, plain forests, mangrove forests and aquatic ecosystems get badly affected due to invading alien floral species.

Invasive alien plant species are non-native species that spread and interfere in a new ecosystem by posing a serious threat to the native biodiversity, leading to economic loss. Invasive species don't allow local species to grow and wildlife to move through. A resin like substance that oozes from such alien species makes the soil acidic, preventing the growth of any other plant species. Species like Lantana, that grows extensively, create a mat-like structure leading to degradation and destruction of the biodiversity. As a result the herbivores like Gaur, Chital and Sambar are deprived of their food. This also affects the survival of the carnivores such as Tigers and Panthers, interlinked to the ecological equilibrium.

In the Western Ghats, where vast plantations of Eucalyptus and Wattle were raised in the past by converting the grasslands and shola forests, the original habitat of the Nilgiri Tahr has been devastated. Nowadays the Indian

Bison makes frequent visits to the Kodaikanal town in Tamil Nadu because of the non-availability of food plants due to extensive plantations of alien species.

Many alien species of flora, introduced mostly by the British, have multiplied to a great extent mainly in the Western Ghats. Vast plantations of Eucalyptus, Wattle (Acacia) and Pinus can be seen across the upper slopes of the Nilgiris and Pulney hills interspersed with Lantana camara. Prosopis juliflora, Parthenium hysterophorus and Eupatorium odoratum can be seen on the lower slopes. Most species such as Eucalyptus, Wattle (Acacia), etc., introduced from Australia have become highly invasive.

Wattle, introduced about four centuries ago to create tannin in the Nilgiris have colonised the grasslands extensively and encroached upon the adjoining Shola Forests. Eucalyptus has become a menace and there is widespread discontent among the locals. Planting of this species has been banned from 1987 onwards in the Nilgiris and the people from the plains are also up with arms against raising of such species either by the Forest department or even by any other wood based industry.

In the last few years, the Forest Department has taken measures to stop the invasive species from spreading such as planting native floral species. The removal of Eucalyptus and Wattle is being carried out in the Reserve Forests of the Nilgiris and Kodaikanal. Although the eradication of these alien species may not be possible in a short span of time, if the measures are taken persistently, it is likely that the landscape may turn into its original vegetation in the long run. Intensive planting of the indigenous floral species should also be conducted after removing the alien species. A dedicated and scientific approach is the need of the hour.

Attempts are being made to eradicate the Prosopis juliflora and Lantana camara as well since they compete to establish the natural regeneration of the native species. Prosopis juliflora, which was introduced during the sixties mainly to meet fuel requirement of the rural population has become a menace. Action is being taken to eradicate it from the Reserve Forests and other water bodies.

Ecologists are of the view that the removal should be in a phased manner with subsequent planting of the native floral species in order to improve the biodiversity. Since the alien species exist for the past three to four decades, complete eradication will take not just time but also sufficient funds and manpower.

It is time State and the Central Governments address the situation with the view to increase the biodiversity and maintain the ecological equilibrium.

Rhododendron nilagiricum raised after removal of wattle in Nilgiris

(*Contd.*)

Photos by Aurhor

Removal of Lantana camara inErode Forest Division

https://www.downtoearth.org.in/blog/how-alien-invasive-plant-species-threaten-western-ghats-62294

Wednesday 28 November 2018

20
AMENDED TAMIL NADU FOREST ACT MUST BE CAUTIOUSLY EXERCISED

While forests under private ownership have already been destroyed with cash crops, if the provisions of the Act are further relaxed the state's green cover will face serious threats.

A country should have at least one third of its land area under forest cover to be able to run a sustained economy, but that has been getting more and more difficult with the ever-increasing global warming and climate change. Every state and country is trying to increase the forest cover by implementing various acts and policies.

For particularly Tamil Nadu, the Tamil Nadu Forest Act was enacted in 1882 during the British rule, with subsequent introduction of the Tamil Nadu Preservation of Private Forests Act (TNPPF Act)-1949, Tamil Nadu Hill Areas (Preservation of Trees) Act-1955, Wildlife (Protection) Act-1972, Forest (Conservation) Act-1980, Tamil Nadu Rosewood Trees (Conservation) Act-1994, etc., the Tamil Nadu Forest Department is trying very hard to protect their forests.

But often forest officials are met with challenges and impediments while enforcing the law. Many complaints are filed against forest personnel who evict encroachers or book them for offences under the TNPPF Act. Human rights petitions filed against the forest staff demoralise them and contempt petitions prevent them from doing their jobs. However, these petitions do not, mostly, stand in the court of law, but they still end up causing mental agony.

Despite this, 20.26 per cent of the state's land is covered with forests. The financial aid provided by Japan to increase tree cover outside the forest has also contributed to these results. But the amendments made in the TNPPF Act have diluted its strictness and handicapped the officials in dealing with the offenders effectively.

While assessing the forest cover, those forests owned by private individuals are also taken into account. In order to protect such forests spread over a vast expanse, Tamil Nadu Preservation of Private Forests Act (TNPPF Act) was enacted in 1949. Out of 32 districts, the Act is in force in around 20. As per Section 3 (1) (a) of the Act, "No owner of any forest shall, without the previous sanction of the (Committee) sell, mortgage, lease or otherwise alienate the whole or any portion of the forest."

There are lot of private forests situated well within and close to the reserved forests of Kanyakumari district. This district was with the Kerala state when the Act was made and it was later introduced here in 1979. Many of the private forests here have been notified under the Preservation of Private Forests Act. Whenever the private estate owners indulged in any act likely to denude the forest or diminish its utility as a forest, immediate action was taken to prosecute them. If the owners tried to develop their land to cultivate commercial crops like rubber, cloves, coconut or areca nut by destroying the existing natural and spontaneous growth, they were booked under the provisions of the TNPPF Act, 1949.

Usually, if a private forest notified under the Act was sold violating the provision under section 3 (1), the matter was reported to the district collector. Then, the sale was declared null and void by the collector. On receipt of the said proceeding, when the aggrieved private estate owner appealed to the state government, it upheld the order passed by the collector. Shocked by the government order, all private estate owners joined together and tried to get the

Act scrapped. The representation made by the association was referred to me, as I was serving as the district forest officer in Kanyakumari, for my remarks.

I said if the Act was lifted, it may lead to denudation of the forest cover from the private estates within no time leading to drying up of the streamlets, streams and rivers which originate from the hills and hillocks on which these private forests are located. Already the district is facing water scarcity during the summer. If any order is passed in favour of the representation, the district may be deprived of not only the water source but also of the valuable biodiversity.

The government did not scrap the Act, but due to repeated pressures and issues raised in the assembly, the state amended the Act sanctioning power to the purchaser under section 4-A (1) Notwithstanding anything contained in sub-section (1) of section 3, the purchaser of the whole or any portion of the forest, which has been sold by the owner of such forest without the previous sanction of the committee for sanction to retain the whole or any portion of the forest, within such time as may be prescribed, etc., and got the assent of the President in 2015.

Certain genuine hardships the private estate owners faced have been taken into account and suitable amendment is issued. But at the same time, unless the provision is used scrupulously, it may allow sale or transfer of ownership of property, deemed illegal, which the original Act did not permit.

When any permission is issued by the committee for sale or transfer of any private forest located inside the forests without proper and thorough field verification and records, it may lead to regularisation of the government forest under encroachment and the illegal structures such as resorts, etc., built destroying the forest.

If the amendment is followed in a casual manner, all the encroachments and the illegal structures that have come up may become legal. Likewise, some of the villages notified under the TNPPF Act located within the core area of the protected areas which act as buffer zones and many villages notified under the Act located in the buffer zone that act as corridors facilitating the wildlife especially the elephants may turn to be a threat to the denizens of the deep forests.

At this juncture, the historical judgement made by the Supreme Court in August 2018 ordering the closure of the illegal resorts built in the Sigur Plateau, the buffer area lying between the Mudumalai Tiger reserve and the Nilgiris Hills with many villages that act as the prime corridor for migrating wildlife especially the elephants, must be recalled.

While the private forest owners are happy with the amendment, the conservationists think that as many of the private estate owners have already denuded the forest cover for raising cash crops like rubber, clove, tea, coconut, etc., taking advantage of the Act's loopholes, the relaxation of the provision may lead to further destruction of forest cover.

Anyway, thanks to the state government for careful examination of the effects of enforcing the legislation. So, only if the authorities empowered to implement the Act are more cautious, the remaining forest cover can be well protected for the sustenance of not only the present generation but also of the posterity.

Indiscriminate felling of trees in private forests

Photo by Author

http://www.downtoearth.org.in/blog/forests/amended-tamilnadu-forest-act-must-be-cautiously-exercised-62662

Thursday 03 January 2018

21
WHY WE NEED A COASTAL ZONE PROTECTION ACT?

If regulations listed in the CRZ notification are implemented properly, coastal zones and fragile ecosystems can be safeguarded.

Coastal zone is a transition zone between marine and territorial zones. It includes shore ecosystems, wetland ecosystems, mangrove ecosystems, mudflat ecosystems, sea grass ecosystems, salt marsh ecosystems and seaweed ecosystems.

It is assessed that about 4800 billion tonnes of domestic waste and 65 million tonnes of solid waste are dumped annually in the sea. Due to continuous onslaught on the coastal areas, the extent of mangroves, coral reefs and fish breeding gets diminished adversely impacting the livelihood of 200 million people who live along the 7517 kilo metre-long coastline of our country.

Hence, it was decided to introduce a plan of action with an aim towards sustained utilization of the coastal zone. Based on that, the Coastal Regulation Zone (CRZ) Notification was issued in 1991 under the Environmental Protection Act, 1986 by the Ministry of Environment and Forests (MoEF) for regulation of activities in the coastal area of India.

Coastal Regulation Zone (CRZ) consists of coastal land up to 500 metres from the High Tide Line (HTL) and a stage of 100 metres along the banks of creeks, estuaries, backwater and rivers where tidal fluctuations occur.

The coastal areas have been classified into four categories-CRZ-I, CRZ-II, CRZ-III and CRZ-IV in the 1991 notification, which aimed at restricting establishment of industries in these areas.

The ecologically sensitive areas that lie between high and low tide line that are very much essential for maintaining the ecosystems are covered under CRZ-I. Natural gas exploration and salt extraction are permitted in this zone.

The areas up to the shoreline of the coast are notified under CRZ-II. Unauthorised structures are not allowed here. Rural and urban areas which fall outside CRZ-I and CRZ-II are covered under CRZ-III. Only agricultural related activities and public facilities are permitted in this zone. Aquatic areas up to territorial limits are notified under CRZ-IV. Though originally this zone was notified for Andaman & Nicobar and Lakshadweep islands, fishing and allied activities were subsequently allowed here after due modification. It's permitted to let off solid waste in this area.

The 1991 notification was amended about 25 times-considering the requests made by state governments, central ministries and NGOs, besides many office orders were issued by Ministry of Environment and Forests clarifying certain issues.

Finally consolidating above modifications, a new notification was issued in 2011 based on the recommendations made by the committee chaired by Dr. M.S.Swaminathan on coastal regulation.

The 2011 CRZ Notification aimed at ensuring livelihood security of the fishing communities as well as other local communities who inhabit the coastal areas, conserving and protecting coastal areas and promoting development in a sustainable manner based on scientific principles.

CRZ area was classified as CRZ-I (ecologically Sensitive areas), CRZ-II (built-up areas), CRZ-III (rural areas) and CRZ-IV (water areas). The only change made was the inclusion of CRZ-IV, which includes the water areas up to the territorial waters and the tidal influenced water bodies.

A separate draft Island Protection Zone Notification was issued for protection of the islands of Andaman & Nicobar and Lakshadweep.

Presently the Ministry of Environment, Forest and Climate Change (MoEFCC) has issued the Draft Coastal Regulation Zone Notification, 2018, based on the representations received from different coastal states, Union Territories and other stake holders in supersession of the Coastal Regulation Zone Notification 2011.

It is clearly mentioned in the Draft that the present notification is issued with a view to conserve and protect the unique environment of coastal stretches and marine areas, besides livelihood security to communities and to promote sustainable development based on scientific principles taking into account the dangers of natural hazards and sea level rise due to global warming.

According to the new notification, the projects which are located in CRZ-I (ecologically Sensitive Areas) and CRZ-IV (areas covered between LTL and 12 nautical miles seaward) will require necessary clearance from the Union government.

The powers for clearances with respect to CRZ-II (areas that have been developed up to or close to the shoreline) and CRZ-III (areas that are relatively undisturbed) have been delegated to the state governments. The construction norms on floor space index (FSI) have been relaxed now. The new notification relaxed the No Development Zone (NDZ) criteria as well.

The Draft notification permits temporary tourism facilities such as shacks, toilet blocks, changing rooms, drinking water facilities, etc., on beaches within 10 metres of the waterline, making the state and even the town planning authorities empowered to grant permission.

Temporary tourism facilities are permitted in NDZ of the CRZ-III areas. Again, the CRZ-I is further classified into CRZ-I A consisting of the ecologically sensitive areas and CRZ-I B covering the area between Low Tide Line (LTL) and High Tide Line (HTL).

After classifying the ecologically sensitive areas under CRZ-I A, activities such as mangrove walks, tree huts, nature trails, etc., were exempted in the name of ecotourism as tourism facilities.

Rural areas with a population density of 2,161/square kilometre, which fall under CRZ-III A shall now have NDZ of 50 metres from the HTL, against 200 metres stipulated in the 2011 Notification.

In the mangrove buffer, laying of pipelines, transmission lines, construction of road on stilts, etc., that are required for public utilities are permitted.

When such activities are permitted in the fragile ecosystems, it is likely to disturb marine life further leading to depletion and destruction of the ecosystems in due course of time.

Developing beach tourism may further lead to conflict with the fisher folk who depend on the beach for their livelihoods. Already such conflict has started at the famous Marina Beach in Chennai's Nochikuppam-a historical fishing village.

The existing service road there was converted into a high-speed concrete road, where the fishermen would earlier sell fish and repair their fishing nets. Setting up of treatment facilities to address pollution is now permissible in CRZ-I B areas. Defence and strategic projects have been exempted.

Cyclone Ockhi that wrecked havoc in Kanyakumari district of Tamil Nadu, the erratic monsoon that brought catastrophe to Kerala, the cyclone Gaja that devastated about 12 districts in Tamil Nadu and the failure of monsoon in the state all show how vulnerable India is to sea-level rise, coastal flooding and climate change.

If the regulations listed in the CRZ Notification are implemented properly, the coastal zones can be safeguarded against encroachments. However, since it is only a notification without any punitive measures, it could not be enforced strictly.

Even Comptroller and Auditor General (CAG) of India pointed out that the frequent amendments, made to the notification, have paved way for commercial and industrial expansion in coastal areas, while natural disasters have become more frequent causing severe loss of human lives and property.

The CRZ Notification 2011 entailed a Coastal Zone Management Plan (CZMP), to be prepared by the coastal states within a year. However, till 2018, many states have not prepared any plan; while few plans submitted lacked any proposal for housing facilities for fisherfolk.

It is high time that the present notification be revoked for the sake of protecting the fragile coastal ecosystems. Promulgation and enactment of a new Act for protection of the coastal zones-with clear classification of various zones, after due consultations with the fishing communities, stake holders, scientists and the department concerned- is the need of the hour.

Courtesy: Google

Violation of CRZ Notification

http://www.downtoearth.org.in/blog/environment/why-we-need-a-coastal-zone-protection-act-62876 Friday 18 January 2018

22

RAGING BULL: HOW THE CONTROVERSY OVER JALLIKATTU ENDED UP SAVING THE PULIKULAM BREED

The ban over Jallikattu and its subsequent lifting has made breeders in Tamil Nadu more conscious about conserving the cattle breed used in the sport, the Pulikulam.

Out of India's 37 indigenous cattle breeds of India, a large number come from Tamil Nadu. These include the Kangeyam, Thiruchengodu, Bargur, Palamalai, Alambadi, Kollimalai, Vadakarai, Manapparai, Umbalachery, Irucchali, Pulikulam, Thambiran madu, Thenpandi, Thondainadu, Thurinjithalai and Punganur.

Out of these Pulikulam breed, native of the Sivagangai district, is mainly used for jallikattu in Tamil Nadu. 'Jallikattu', otherwise known as 'Eru Thazhuvuthal' (Tamil for 'Bull Embracing') is a sport that has been played in southern Tamil Nadu during Thai Pongal since the Sangam Period.

With the advent of green revolution, due to farm mechanisation, the use of draught animals has decreased drastically. According to the last Livestock Census of India, while the indigenous cattle population has decreased considerably, exotic and crossbred cattle have increased. Indigenous breeds and hybrids with dominant native genes produce A2 milk, which is good for human health. Jallikatu is considered to be a bio-cultural sport as it helps to conserve native cattle breeds. But, fortunately or unfortunately the Supreme Court of India banned Jallikattu on May 7, 2014 based on a plea by the Animal Welfare Board of India (AWBI) and People for the Ethical Treatment of Animals (PETA).

I personally feel that the ban introduced was also a blessing in disguise. That is because in the past, when the event was organised without proper regulations, many young men who participated in the bull fight had lost their lives leaving their wives and children widowed and orphaned. The participating bulls were tortured by biting their tails and fed with alcohol to make them more ferocious. The bull tamers were also in the habit of consuming alcohol before joining the event.

But, the massive, unprecedented and peaceful protest organised by the people of Tamil Nadu, referred to as the 'Jallikattu Uprising' in support of Jallikattu forced the government of India to clear the ordinance proposed by the Tamil Nadu Government and a suitable order was issued during January, 2017, as an attempt to preserve the cultural heritage of Tamil Nadu and ensure the survival and well-being of native breeds of bulls. Subsequent to the order permitting the sport, many restrictions were imposed for safe and smooth conduct of the bull fights. Medical check-ups for both, the participating men and bulls were made compulsory, with necessary dos and don'ts to avoid any untoward incidents. Otherwise if the sport is banned permanently, the Pulikulam breed also may vanish gradually.

The sale prices of this breed plummeted to an all-time low in and around Madurai because of the ban on Jallikattu. When there was ban on the sport, the condition of the Pulikulam was pathetic. People were not ready to accept any newborn calf even free of cost. Instead, they started demanding money from the owner of the animals. Famers started selling the bulls to slaughter houses of Kerala. But, when once due permission was granted by both governments, the entire situation changed. Those who were interested in rearing Pulikulam bulls started bribing cowherds to inform about the delivery of a male calf after noticing any pregnant cow in the herd. A newborn Pulikulam calf fetches around Rs.10,000 per head.

The staging of cattle fairs helps farmers to identify viable stock which ensures a vigorous progeny. We can have sustainable agriculture only through native cattle breeds as they help in maintaining pest control and soil health and milk yield with minimum maintenance cost. Besides, they are resistant to climate change.

Ignoring the quantity, when the quality is taken into account, the milk produced is said to be very good besides double the number of calving by the native breeds.

The breed derives its name from the village from where it originates: Pulikulam in Sivagangai district. It is said that this village derived its name 'Pulikulam' (Tiger Pond) a few centuries back, because of the presence of tigers in the dense forests that used to quench their thirst in the pond. It is said that these bulls were capable of fighting ferociously with tigers. It is also bred in Madurai, Virudunagar and Theni districts in Tamil Nadu. 'Palingu maadu', 'Mani maadu', 'Jallikattu maadu', 'Mattu maadu' and 'Kilakattu maaadu' are the other names of this breed. This indigenous breed is popularly used for Jallikattu. It is a drought-resistant breed and is used as a draught breed rather than milk production as the yield is less compared to other breeds. It supplies good manure and muscle power for ploughing the land, thus contributing to significant organic farming.

The Pulikulam's population of around 90,000 in 1995 has come down drastically to about 16000 animals in recent years in and around Madurai district. In the past, while Yadava Community made up 99% of the cowherds rearing the Pulikulam, while the remaining 1% was from the Mukkulathor community. But now the trend has changed. Many people of different communities rear this breed for the sake of Jallikattu as it is considered to be a prestigious sport.

In the past this breed was maintained in the villages from October up to the harvest season in January and then would let loose in the nearby hills and forests for grazing. But, the restrictions by the forest department for grazing and penning inside the forests led to the dwindling of their numbers. Today these animals are penned overnight in the fields of private individuals where their urine and dung fertilises the soil. Farmers from Kerala pay Rs.10 for a kilogram of this breed's dung, which is used as manure for cash crops such as pepper and cardamom.

This breed is resistant to foot-and-mouth disease, tuberculosis and brucellosis. Generally the males will be dark grey in colour while the females will be white or grey. While the males have large humps, the females have small ones. They are very active with ferocious nature and power of endurance. The bulls are very vigorous, strong and swift. They are aggressive and have the power of endurance. The price of a 6-month-old male calf is Rs.20,000-25,000. If a mature bull is undefeated in Jallikattu, its price can go up to Rs.5 lakh.

The Breed Registration Committee (BRC) of National Bureau of Animal Genetic Resources (NBAGR) under the control of the Indian Council of Agricultural Research (ICAR) has approved and registered Pulikulam breed as an indigenous breed with the accession number '01 Pulikulam Tamil Nadu INDIA_CATTLE_1800_PULIKULAM_03035.

The future

Jallikatttu events were organised only in a few places in and around Madurai in the past. But after the ordinance permitting the event was passed in 2017, in many parts of the state, Jallikattu is now organised in a grand manner, with proper procedure, after getting due permission from the district collector concerned.

The mega Jallikattu organised in Viralimalai of Pudukottai district on January 20, 2019, with 1,353 bulls and around 500 tamers, has created world record. Though medical centres and emergency operation theatres were arranged in advance to treat persons with simple as well as grievous injuries, two persons were gored to death. Fortunately insurance has now been introduced for the first time not only for the tamers and the bulls, but also for the spectators. Valuable prizes distributed to the best bull-tamers and the owners of the best bulls include two cars, 10 motorcycles, 700 bicycles, gold and silver coins and home appliances.

Thus the lifting of the ban on Jallikattu, a historical sport considered to be a sign of bravery in Tamil Nadu, has helped the survival and betterment of the rare Pulikulam breed.

Jallikattu organised at Viralimalai, Trichy District with a world record of 1353 bulls and 500 Tamers on 20 January, 2019

(Contd.)

Courtesy: Google

Puikulam bull

https://www.downtoearth.org.in/blog/wildlife&biodiversity/raging-bull-how-the-controversy-over-jallikattu-ended-up-saving-the-pulikulam-breed-62900

Monday 21 January 2019

23
HOW TO SAVE INDIA'S ELEPHANTS FROM KILLER RAIL TRACKS

The government, railways and forest department have it within their power to initiate a number of measures to reduce elephant deaths on rail tracks.

The extensive network of Indian Railways cut across dense forests, habitat of different wild species. In many states, railway lines passing through elephant habitats have led to numerous accidents and the death of 249 elephants during 1987-2018.

Between 2015 and 2018, 49 elephants have been killed in train accidents (9 in 2015-16, 21 in 2016-17 and 19 in 2017-18). Of these Assam and West Bengal accounted for 37 deaths and Tamil Nadu, Odisha, Andhra Pradesh and other states for the remaining. While the numbers have gradually decreased from 2015-16 to 2017-18 for West Bengal, in Assam were on the rise.

The Union Ministry of Environment, Forest and Climate Change (MoEF & CC) acted as early as in 2016 to mitigate human-elephant conflict, but elephants have continued to die due to speeding trains. This despite the MoEF & CC declaring the elephant a 'National Heritage Animal' in 2010, considering the valuable ecological services rendered by the species.

Elephants are architects of the forest and woodland ecosystem. Many ecologists feel woodlands may cease to exist without elephants. In tropical forests, 30 percent of the gigantic tree species and 40 percent of the tall tree species depend on elephants for seed dispersal. Considered nature's 'gardener', they are key in shaping the landscape, in pollination, germination of seeds and improving the fertility of forest soil with heaps of dung.

Given the importance of the elephant, urgent measures to reduce its mortality are to be thought over seriously, discussed in detail and implemented properly.

Why the accidents?

In a first, the Divisional Forest Officer of Walayar in Kerala booked a train driver for hitting and knocking an elephant 100 metres away while speeding in July 2016 under Section 9 (Hunting and the intent to hunt) of the Wildlife protection Act, 1972. However, I feel that mutual understanding and friendly coordination between the forest department and the railways are very much required to sort out the issue.

The importance of elephants for the existence of the forests-which provide for humanity's basic needs, including pure air, clean water, medicine and energy and also ensure food security-should be explained to railway officials. They should be taken deep inside the forest area for this. Such sensitization programmes, especially for train drivers, guards and station masters may be of great help in averting any kind of possible tragic events. If needed, the frontline forest staff also may be allowed to travel along with the railway engine driver to identify the vulnerable spots so as to ensure the safety of the animals.

At the same time, forest officials also should understand the difficulties faced by the railway personnel while driving fast-moving trains through dense forests, especially during night hours.

In certain places, when the elephant tries to cross the railway track, it gets trapped between steep embankments on either side of the track. In such cases, the steep embankments are to be flattened and periodical patrolling is very much required.

Due to the obstruction of the view of the railway track, especially at bends, caused by the vegetation that has grown on either side of the track, certain unavoidable accidents take place. In order to avoid such incidents, clearing of the vegetation all along the track should be carried out periodically. Solar-powered lights at curves and cuttings can be provided to have better view for a long distance.

The accidents have increased after gauge conversion and also due to enhanced speed of the train from 60 kph to 100 kph. Of course, the increase in the elephant population has also contributed to the increasing number of accidents. In addition to elephants, other animals like gaur, sambar, leopard and tiger have been killed colliding with trains due to the expansion of the railway network, gauge conversion, over-speeding and increased frequency.

What can be done?

When a study was undertaken on the frequent accidents caused due to elephants colliding with fast-moving trains, it was found that most of the accidents have taken place during the night- from 6 pm to 6 am-rather than day as elephants rest in dense forest during day time. Reducing the operation of trains during the night to the barest minimum is one way to avoid this situation. Railway authorities should see that the speed of the trains that pass through forests in which wild animals reside, is regulated between sunset and sunrise. A minimum speed of 30 kilo metres/hour is to be maintained when trains pass through dense forests at night. Utmost care should be taken at vulnerable sections.

At places where elephant corridors and railway tracks intersect, the construction of underpasses or overpasses can be planned enabling the animals to cross over without any difficulty. A study has shown that ramps constructed near rail tracks in Coimbatore division of Tamil Nadu have helped the pachyderms to return to the forest when they noticed a fast moving train. These ramps have been planned based on the camera trap information and built meticulously as a protective measure.

Warning signboards indicating the movement of animals across the tracks can be erected at vulnerable stretches in order to alert train drivers.

Railway staff can be deputed for keeping the track free from any food waste as the same may attract the wild animals. Train passengers should be made aware about not disposing of waste food and other eatables on the tracks. Coordinated patrolling of the vulnerable tracks by forest, wildlife and railway staff can be organised especially during the night time and during crop-raiding seasons and information about the presence of elephants can be passed on to the railway signalmen. As patrolling inside thick forest at night may be dangerous, more watch towers need to be erected at vulnerable spots.

Electronic surveillance equipments, cameras and wireless sensors should be installed along accident- prone stretches so as to avoid any accident. Infrared beams can be used for detecting elephant movements and acoustic devices to scare them away from the tracks.

Vulnerable stretches, which are known for frequent and large number of accidents, can be identified and tracks there can be realigned.

Excluding the sections of track used by elephants to cross, the remaining parts should be fenced and elephant-proof trenches dug out, with the idea of preventing them from crossing the vulnerable stretches. During summer, elephants are in the habit of roaming over long distances in search of water and food. In order to curtail their movements across the tracks, ponds can be created on the sides in addition to developing fodder resources in the forest areas.

A map can be prepared showing the sites of train accidents in which animals died, the time, periods of collisions (many collisions have taken place during the paddy and maize harvesting seasons), elephant crossing points and the routes of the movement of the animals along the tracks, and exhibited at the concerned railway stations and forest offices, with the idea of avoiding any possible accidents by creating awareness on the movement of the elephants.

All the measures must involve the Indian Railways, MoEF &CC, forest departments of the concerned states, non-profits, biologists, engineers, technocrats, estate managers, conservationists, ecologists and locals who live near such areas, to evolve some effective methods to save the elephants from the fast moving trains.

The Tamil Nadu experience

The Tamil Nadu forest department has installed infrared sensors on 6 metre-poles and on trees on either side of the railway track close to the elephant corridor in the Coimbatore forest division. If the sensor senses any elephant, a sound alarm is set off and alerts forest officials, enabling them to rush immediately to the spot to drive away the animals. Also, a text message is sent to the concerned railway control room to pass on the message to the train driver in order to reduce the speed.

Officials of the Eastern Railway have erected devices which loudly broadcast the buzzing sound of honeybees in order to drive away the elephants, as they are scared of the insects' buzzing noise.

The forest department of Uttarakhand state has started using drones to track the elephant movement.

Felicitation of train drivers and other officials, who avert any possible collision, can boost the morale and motivate them to improve their efficiency towards saving the pachyderms.

Adequate and timely compensation for crop damage, grievous injury or human casualty should be arranged and provided to affected families.

The capabilities of the field personnel of the forest department as well as of the railways should be enhanced through periodical workshops and field visits by experts on wildlife and biodiversity.

In the past, wastelands stretched from the boundaries of reserve forests to human habitations, with scattered tree growth. These acted as the corridors for wild animals, especially elephants. The owners of these lands raised rain-fed crops such as jowar, foxtail millet, kodo millet and little millet, only for a few months in a year. The elephants moved along the migratory routes from time immemorial without causing any trouble to anyone.

Currently, farmers have started raising crops like sugarcane, banana, maize and areca nut using borewells and tubewells, with electric fences all around these lands. Attracted by the palatable crops, elephants get electrocuted while they touch the electric fence. In places without any fence, elephants raid crops. Farmers can be advised to cultivate alternate crops like castor, chilli, cotton, gingelly, onion, garlic, ginger, mulberry, tobacco and turmeric, which are not preferred by elephants.

Besides, most wastelands have been converted into real estate, farm houses, mystic centres, resorts and educational institutions, without getting any permission from the authorities. As a result, elephants have been forced to move towards railway lines.

In Tamil Nadu for instance, most dry lands, though owned by private individuals, have been notified under the Tamil Nadu Preservation of Private Forest Act (TNPPF Act) and Hill Area Conservation Authority (HACA). To be frank, the concerned government departments and the revenue officials were not aware of the aftermath of the changes in crop patterns and other developments taking place in these areas, resulting in human-animal conflict. Consequently, when there was a hue and cry about human-animal conflict, the district administration took serious action to shut down educational institutions and other resorts in Coimbatore district near Mettupalayam at the foothill of Nilgiris hill range and Sathyamangalam Tiger Reserve (STR). Of course, the intervention of the Supreme Court of India in saving the elephants by ordering the closure of the resorts along the Sigur plateau lying between Mudumalai Tiger Reserve (MTR) and the Nilgiris hill range is known to everyone.

Considering the above facts, if serious and earnest efforts are taken at appropriate levels, the killing of elephants by speeding trains can be reduced to a great extent, thereby mitigating the human-elephant conflict considerably.

The author is President, the Society for Conservation of Nature, Trichy, Tamil Nadu and consultant with the Society for Social Forestry Research & Development, Tamil Nadu.

Courtesy: Google

Elephant mowed down by a speeding train near Madukarai in Coimbatore District

https://www.downtoearth.org.in/blog/wildlife-biodiversity-how-to-save-india-s-elephants-from-killer-rail-tracks-63126

06-February-2019

24
IMPLEMENT THE BIOLOGICAL DIVERSITY ACT IN ITS TRUE SPIRIT

It is high time the Centre and states did so. It would save India from a number of avoidable problems.

Biodiversity is very crucial for the functioning of ecosystems that provide us with various products like oxygen, food, fresh water, fertile soil and fuel, besides ecological services such as moderating storms, mitigating climate change, making it possible for millions of species, including humans to survive on earth.

According to the International Union for Conservation of Nature (IUCN), 17,291 species out of 47,677 are 'threatened' with extinction. Seventy nine mammalian species out of 5,490 are 'extinct', with 188 'critically endangered', 449 'endangered' and 505 'vulnerable'. A total of 1,895 out of 6,285 amphibian species are in danger of becoming extinct. As per an IUCN report, the abundance of species has declined by 40 percent between 1970 and 2000.

Alongside species extinction, habitats are being destroyed, with lands being converted indiscriminately for development; there is also spread of invasive alien species, leading to climate change and pollution. Six million hectares of forest have been lost annually from 2000.

India is home to the Western Ghats and the Northeast Himalayas, two important biodiversity hotspots out of the world's 25. The country stands alongside 16 other mega-biodiversity countries bestowed with 7-8% of the world's species.

On June 5, 1992, India signed the Convention on Biological Diversity at Rio de Janeiro which provides a framework for the sustainable management and conservation of our country's natural resources.

Ten years later, the Biological Diversity Act was enacted in 2002 in order to conserve biodiversity, manage its sustainable use and enable fair and equitable sharing benefits arising out of the use of biological resources with the local communities.

Though 17 years have passed since the enactment of the Act, most of the local bodies of 23 states have not prepared the People's Biodiversity Registers (PBRs), which are considered to be the basic records of a region's biological resources such as plants, animals and the traditional knowledge of the local people.

In the absence of the PBRs, when the Act is not implemented, whatever environmental clearances given for various projects without recording the real state of biodiversity, results in the environment impact assessment reports becoming illogical and invalid.

Under the Act, National Biodiversity Authority (NBA) has been formed and is headquartered at Chennai. State Biodiversity Boards (SBBs) also have been formed in all Indian states. But the constitution of the Biodiversity Management Committees (BMCs) at local levels-in panchayats, municipalities and corporations in many states- is pending, causing delay in preparation of PBRs of bio-resources (both wild and cultivated).

In April 2004, the Ministry of Environment and Forests (MoEF) notified the Biological Diversity Rules 2004 under the Biological Diversity Act, 2002. The Act has a unique system of governing access and benefit sharing (ABS) through NBA, SBBs and BMCs formed at different levels.

As the Act provides measures for safeguarding traditional knowledge, preservation of threatened species and prevention of bio-piracy, many states have come forward to implement it in its true spirit.

Kerala was the first state to form BMCs in 978 village panchayats, 60 municipalities and five corporations in the state and the preparation of PBRs have also been completed. Since the then Chief Minister Ommen Chandy of Kerala showed real initiative, it successfully followed all processes with the view to protecting and conserving the state's bio-resources.

In neighbouring Tamil Nadu, due to delay in conducting the local body elections, works concerned with formation of BMCs and preparation of PBRs have not been completed, resulting in the gradual depletion and pollution of the natural resources.

Proper implementation of the Biological Diversity Act

As the Biological Diversity Act is not given due importance, protection of forests and wildlife has become secondary in many parts of India, leading to water scarcity and man-animal conflict. Reliable sources disclose that less than 3 percent of local bodies, spread over 15 states had prepared PBRs till 2018.

Every year, the 'International Day of Biodiversity' is observed on May 22 in most government departments, educational institutions and other conservation-oriented organisations. In fact, the year 2010 was observed as the 'International year of Biodiversity' without realizing the importance of implementing the Act.

Neither the central nor state governments appear to be bothered about mitigating anthropogenic disasters by implementing the Biological Diversity Act, which is supposed to provide suitable remedial measures for all human-made evils.

Most major Indian rivers like the Ganga, Yamuna, Gomti, Chambal, Mahi, Vardha, Godavari, Damodar, Sabarmati and the Cauvery have been badly polluted due to dumping of untreated sewage, chemical wastes and industrial pollutants. But, unfortunately, the same water is being used for drinking, bathing and washing. If the Act is enforced properly, it may definitely help in reducing air and water pollution to a great extent.

During 2014, the number of threatened species was 988. But in 2017, the number has increased to 1,065 with an increase of 7.8 per cent when compared with the previous assessment according to the 'Red List' of threatened species published by IUCN. If the Act has been implemented on time, this increase could have been prevented or at least reduced to a great extent.

Experts are of the opinion that the country is losing minimum of Rs.30,000 crores annually by not implementing the Biological Diversity Act, which has provisions for access and benefit sharing for commercial utilization of bio-resources.

It is ascertained that the NBA headquartered at Chennai is losing about Rs.10,000 crores annually from foreign companies as the guidelines are not applied properly. If the money had been realised through implementation of the Act, it would have been utilized for conservation of forests, wildlife, ecosystems and other bio-resources. As the Ministry of Environment, Forest and Climate Change (MoEF & CC) is the authority to implement the act, it should see that the provisions of the Act are enforced without any delay for conserving and sustainable management of the biodiversity.

It is high time that the central and state governments should wake up from the deep slumber and take speedy and earnest action to overcome the disastrous calamities by implementing the Biological Diversity Act in its true spirit.

Credit: Getty images

The NGT suspended the permission granted for construction of hydro-electric project in Zemithang Valley of Arunachal Pradesh as it is the wintering site of the endangered Black-necked Cranes

https://www.downtoearth.org.in/blog/wildlife-biodiversity/implement-the-biological-diversity-act-in-its-true-spirit-63322 Thursday 21, February 2019

25
EASTERN GHATS: A BIOTA UNDER SERIOUS THREAT

An effective strategy for eco-development, involving locals can save the flora, fauna and natural resources of the region.

The Eastern Ghats once covered by luxuriant forests are becoming barren because of the greed and endless quest of mankind in the name of development. The series of broken hills run from the Mahanadhi in Odisha to the Vaigai in Tamil Nadu, with an average elevation of 1000 metres in peninsular India.

In Tamil Nadu, the Ghats comprise the Javadhi hills, Yelagiri, Balamalai, Bargur, Servarayan, Bodhamalai, Chitheri, Kalrayans, Kolli hills, Pacha malai, Piran malai, Semmalai, Sirumalai, Karanthamalai, Azhagar malai, etc. The hills run west and south western to merge with the Western Ghats near Doddabetta in the Nilgiris.

They have a rich biota and a fragile ecosystem, which is being degraded by illegal logging and exploitation of forest wealth. Also, tribes living in these hills for ages, have been exploited by wealthy and politically strong mafia.

There is an urgent need for the Tamil Nadu government to intervene. It should take effective and earnest action to save the major biosphere from ruthless hands.

These hills get an average 1000 millimetres of rainfall, mostly from the north-east monsoon. Rivers like Naga nadhi, Cheiyaru, Amirthiyaru,

Kovilmalaiyaru, Idiyaru, Mriganda nadhi, Pennaiyaaru, Puliyancholaiaaru, Aiyaaru, Swetha nadhi, Cauvery, Sarabanga nadhi, Thirumanimuthaaru, etc., originate from the hills of the Eastern Ghats. Their water irrigate Lakhs of hectares at the foot hills and the nearby plains.

Waterfalls such as Periyar falls, Megam falls, Beeman falls, Killiyur falls, Hogenakkal falls, Agaya gangai, Jalagam parai falls, etc., make the land fertile apart from drawing tourists.

The rivers have dams like Krishnagiri, Sathanur, Shenbagathoppu, Kuppanatham, Varattaaru-Vallimadurai, Kariyalur, Gomuki, Andiyappanur, Mriganda, Chengam, etc. They help in irrigation by controlling water flow and also serve as picnic spots.

Not known as widely as the Western Ghats, the Eastern Ghats biospheres are also major ones, with rich forests, perennial and semi-perennial streams and other natural resources.

The forests include dry evergreen forests, semi-evergreen forests, southern tropical dry mixed deciduous forests, dry savannah forests, southern tropical dry scrub forests, southern tropical thorn forests, Carnatic umbrella thorn forests, southern sub-tropical hill forests and southern thorn scrub.

Valuable trees like Eetti (Dalbergia latifolia), Semmaram (Pterocarpus santalinus), Vengai (Pterocarpus marsupium), Pala (Artocarpus heterophyllus), etc., are present here in addition to rare and endangered species such as Mara uri (Antiaris taxicaria), Hildegardia populifolia, etc.

Mara uri was thought to be restricted to the Western Ghats, but has been found in Pachamalai too. Similarly Hildegardia populifolia-on International Union for Conservation of Nature's Red List (for critically endangered species) was earlier reported only from Kodaikanal, but has been found in Kalrayan Hills also.

Though Mahogany (Swietenia mahogany) grows mostly in moist regions, it has been around in Pachamalai-a proof of the richness of the ecosystem.

The Eastern Ghats ranges are home to 2,500 flowering plants, thus protecting 13 per cent of India's flowering plants. They are also the habitats for wild animals such as elephants, panthers, the Indian bison, bears, deer, wild boar, slender loris, mongoose, jungle cat, wild dogs, porcupine, hare, toddy cat, monkey and reptiles such as python, monitor lizard, etc. About 290 bird species and nearly 4,000 insects are also found.

The tribes living in the hills have not received proper price for the food grain they cultivate. They have thus switched over to monocropping tapioca, leading to deterioration of land.

Most Reserve Forest areas in these hills are controlled by the forest department, protected by the Tamil Nadu Forest Act, 1882 and Wildlife Protection Act, 1972. The Tami Nadu Hill Areas (Preservation of Trees) Act, 1955 is in force in certain parts of these hills, but it is not implemented effectively in the absence of popular support. As a result, the protection of the tress in private holdings has become a challenge.

By making use of the loop holes in the Act and other government rules, the trees are felled and removed clandestinely. These activities have made the hills of the Eastern Ghats barren, its streams have run dry and the biodiversity is disappearing gradually

Large scale plantations of coffee, tea and orchards have been raised in these hills. Silver oak trees grown in these plantations as shade trees also are removed gradually, weakening the fragile ecosystem. While the plantations of coffee and tea are considered to be the cause of degradation in places like Yercaud hills, monocropping in Kolli hills and Pachamalai has devastated entire hills, depleting the native vegetation.

The aromatic and most valuable sandal trees, once growing abundantly in Chitheri, Javadhis, Yercaud, Kolli hills, Pachamalai, Bodha malai, etc., have been indiscriminately felled and illegally removed. The precious species has almost vanished. Only younger regeneration is noticed in a few pockets in some hill ranges.

Even the rare medicinal plants like Sirukurinjan (Gymnema sylvestre), Milagu (Piper nigrum), Kattu kodi (Smilax zeylanica), Vasambu (Acorus calamus), Kanthal (Gloriosa superba), etc, face extinction because of uncontrolled removal.

Non-native species such as poochedi (Lantana camara), vengaya thamarai (Eichhornia crassipes) and veli karuvai (Prosopis juliflora) have become invasive, destroying native species and leading to ecological imbalance.

The removal of enormous quantities of bauxite and magnesite ore from Kolli hills and Servarayan hills, respectively, led to indiscriminate destruction of forests. Consequently, water resources in these regions have dried up.

The Supreme Court, fortunately, banned removal of minerals from these hills, once rainfall became erratic.

The forest department permits the removal of Kadukkaai, Nellikai, Mahali Kizhangu, Kalakkai, Eenjamaaru, Kattu Karuveppilai, Kattu mango, etc, as minor forest produce. As a result, the availability of food for the wild animals is scarce.

Despite the Wildlife Protection Act, hunting takes place in some pockets. Tribal villages, private lands and private estates have fragmented protected forests. The forests are getting degraded because of the illicit collection of fire wood, illicit grazing and illicit felling of trees.

Annual forest fires have become a serious cause for the loss of biodiversity. The pollution caused by the industries established near forest areas also poses a serious threat.

Indiscriminate destruction of the forests has increased human-animal conflict in recent years. Water scarcity and a threat to habitats drive animals to cultivated lands and human habitations in search of food and water.

The pristine environment of Yercaud, Yelagiri, Javadhis, Kolli hills, Kalrayans, Sirumalai, Thiruvannamalai, Hogenakkal, Azhagar malai, etc., are deteriorating due to un-regulated tourism. Un-treated sewage and plastic waste have degraded the environment. Only if the people realise the importance of a clean environment and cooperate with the local administration, the natural resources can be protected.

The concept of eco-tourism introduced by the Forest Department in recent years involving local forest stake-holders may be a great boon to not only residents but also the biota of the entire region.

While eco-tourism programme takes care of the livelihood of locals by making them stake-holders, the forest and its biodiversity are guarded by the eco-tourism management committee members in turn. The experience has been positive in Srivilliputhur Grizzled Squirrel Wildlife Sanctuary.

Only if an effective strategy for the eco-development of these regions is developed by involving the local people, the flora, fauna and other natural resources of Eastern Ghats can be protected. Otherwise the future of the Eastern Ghats and its precious biodiversity will be only an illusion.

Photo by Author

View of Pachamalai Hills

https:www.downtoearth.org.in/blog/wildlife-biodiversity/eastern-ghats-a-biota-under-serious-threat-63456.

March 6, 2019.

26

PROPAGATE AND PROTECT HALOPHYTES

It is high time that India's governments make serious efforts to identify, propagate and protect these plants which establish mangroves, provide nutrition and even mitigate climate change

Halophytes are salt-tolerant plants that grow in waters with high salinity, such as in mangrove swamps, marshes, seashores and saline semi-deserts. Only two per cent of the plant species found on the Earth are halophytes. As they are able to tolerate high salinity through different adaptation methods like tolerance, resistance and avoidance, they have less competition in saline environments.

Credit: Getty Images

Halophytes help in regenerating Mangrove forests

Halophytes are classified into aqua-halines, terrestro-halines and aero-halines.

Aqua-halines include Emerged halophytes, among which most of the stem remains above the water level and Hydro-halophytes among which almost the whole plant remains under water.

Terrestro-halines include Hygro-halophytes, which grow on swamp lands, Meso-halophytes, which grow on non-swamp and non-dry lands and Xero-halophytes, which grow on dry or mostly dry lands.

Aero-halines include Oligo-halophytes, which grow in soil with NaCl (Sodium Chloride) from 0.01 to 0.1 per cent; Meso-halophytes, which grow in soil with NaCl from 0.1 to 1 per cent and Euhalophytes that grow in soil with NaCl greater than one per cent.

In the state of Tamil Nadu, where I come from, halophytes are found growing in waters with high salinity along the seashore in Pichavaram in North Tamil Nadu as well as the Point Calimere Wildlife and Bird Sanctuary and Gulf of Mannar Biosphere Reserve along the Palk Strait in the southern part of the state. These are the associates of the true mangrove plants.

Some of the mangrove associates that grow abundantly in Point Calimere Wildlife and Bird Sanctuty and Muthupet areas include Suaeda monica, Suaeda maritime, Suaeda nudiflora, Arthrocnemum glaucum, Arthrocnemum indicum, Salicornia brachiata and Sesuvium portulacastrum.

Out of these halophytes, plants belonging to the Suaeda species have been growing in large numbers. They are herbaceous plants growing to a height of 1 to 2 metres. Suaeda maritime has red streaks on the main stem which is missing in Suaeda monica.

The Suaeda and Salicornia varieties do not compete with each other. Unlike other plants, those of the Sesuvium genus grow by spreading on the floor. Their stems and the leaves are red in colour.

Halophytes are generally colonizers and grow in areas which are subjected to periodic inundation. Suaeda monica is the only halophyte which is observed growing more gregariously in degraded areas when compared with other halophytes.

As the halophytes are salt extractors, they perform the key function of reducing soil salinity and make the area more suitable for growing mangrove species. Over a period of time, in the course of ecological succession, halophytes are gradually replaced by mangrove species, which grow vigorously and get established.

If not for the presence of halophytes, the regeneration of mangroves would be very difficult in seashore areas which face frequent sea-water ingress.

Sea water generally contains 40 grams of dissolved NaCl per litre. Beans and rice can grow in water containing about 1-3 grams of NaCl/ litre. Barley and Date Palm can grow in water with 5 grams of NaCl/ litre. Though these plants have more salt content, parts of them are used for human consumption and as cattle feed.

Halophytes that grow in desert regions, estuaries where the rivers join the sea and along the seashore, are used for human consumption, cattle feed and for biofuel production. In Point Calimere, the succulent leaves of Suaeda monica are relished by black bucks while in Somalia and in Kenya, it is browsed by camels and goats.

It is said that during famine, the leaves of Suaeda monica were boiled and eaten by the people of this region. Edible oil is extracted from Salicornia species. The salt absorbed through its roots is evaporated through the minute pores of the leaves of the halophytes.

Generally halophytes will have very small leaves. When there is scarcity of water, they shed them. These plants are bestowed with long roots that help in search of fresh water. When the leaves are absent, photosynthesis will take place through the stems.

Some of the halophytes grow on top of the hills situated near seashores and on sand dunes. They are capable of thriving even in semi-desert regions with scanty saline water. The succulent leaves help the halophytes for storing water in these kinds of adverse conditions.

A treasure trove from Nature

Halophytic flora plays a major role in protecting coastal habitats and maintaining ecological stability. They often creep and act as sand-dune binders. They prevent erosion to a great extent and seawater incursion into freshwater habitats. Halophytes provide food and shelter for large number of aquatic and terrestrial animal species.

Researchers have identified some halophytes which have the potential of yielding sustainable supply of renewable resources like food, fodder, fibre, fuel, green manure and raw materials for pharmaceutical, industrial and household products.

Studies have shown that many halophytes can be used as sources of nutritious grain and oil, while some bear edible or economically useful roots, bark, stems, leaves, flowers, fruit and seeds.

Certain halophytes are able to accumulate and transform toxic levels of heavy metals (lead, cadmium and selenium) into organic compounds. Salicornia cultivation is considered to be economically profitable as it is harvested as selenium-rich animal feed because of its ability to convert high concentrations of inorganic selenate into organic selenium. Salicornia plantation has been initiated in Mexico as it plays a key role in carbon sequestration.

The commercial viability of the halophytes is determined by screening relevant properties such as salt tolerance, nutritional value, palatability and digestibility. Halophytes are assessed to have potential as biomass crops to directly sequester up to 0.7 G.T C. Thus, halophytes, like trees, can play a significant role in mitigating climate change.

In reality many of the front line field staff of the forest departments in India are not able to identify these valuable halophytes. The ecological values of the halophytes are not known to many. So far, no tangible action has been initiated to propagate and protect halophytes by any Government agency.

Since halophytes play an important role in helping the establishment of mangroves which in turn protect the interior from tsunamis, cyclones and storms, it is very much essential to propagate and protect them. It is high time that the Central and State Governments make some serious efforts to identify, propagate and protect the most useful halophytes.

The author is President, the Society for Conservation of Nature, Trichy, Tamil Nadu and Consultant with the Society for Social Forestry Research & Development, Tamil Nadu.

---

https://www.downtoearth.org.in/blog/forests/propagate-and-protect-halophytes-63711. March 26, 2019.

27
SCIENTIFIC MANAGEMENT OF MANGROVES IS THE NEED OF THE HOUR

40 percent of mangrove forests in West Coast of India have been converted into farmlands and housing colonies over the last three decades.

Mangroves are salt-tolerant vegetation that grows in intertidal regions of rivers and estuaries. They are referred to as 'tidal forests' and belong to the category of 'tropical wetland rainforest ecosystem'.

Mangrove forests occupy around 2,00,000 square kilometres across the globe in tropical regions of 30 countries. Out of the total mangrove cover of 4,482 sq km in the India, Tamil Nadu has about 23 sq km.

Mangrove ecosystem is the interface between terrestrial forests and aquatic marine ecosystems. The ecosystem includes diversified habitats like mangrove dominant forests, litter laden forest floors, mudflats, coral reefs and contiguous water courses such as river estuaries, bays, inter-tidal waters, channels and backwaters.

Mangroves are trees and shrub species that grow at the interface between land and sea in tropical and subtropical regions of the world, where the plants exist in conditions of salinity, tidal water flow and muddy soil.

The structural complexities of mangrove vegetation create unique environments which provide ecological niches for a wide variety of organisms. Mangroves serve as breeding, feeding and nursery grounds for most of the commercial fishes and crustaceans on which thousands of people depend for their livelihood.

Mangroves give protection to the coastline and minimise the disaster due to cyclones and tsunami. Recent studies have shown that mangroves store more carbon dioxide than most other forests.

Mangroves are intermediate vegetation between land and sea that grow in oxygen deficient waterlogged soils which have Hydrogen Sulphide (H_2S). They perform important ecological functions like nutrient cycling, hydrological regime, coastal protection, fish-fauna production, etc.

Mangroves act as shock absorbers. They reduce high tides and waves and help prevent soil erosion. They also provide livelihood opportunities to coastal communities.

Sundarbans in the Gangetic delta with an area of 2.12 lakh hectares (ha) supports 26 plant species of mangrove with a maximum height of more than 10 metres. Pichavaram in Tamil Nadu with an area of 1,100 ha supports 12 plant species growing to a height of 5 metres.

Muthupet is the biggest mangrove forest in Tamil Nadu with an extent of 11,885.91 ha with poor species diversity due to lack of fresh water supply. But mangroves are being destroyed and facing severe threats due to urbanisation, industrialisation, and discharge of domestic sewage, industrial effluents and pesticides. Saltpans and aquaculture also pose major threat to the mangroves.

As a result, mangroves get depleted to the tune of 2-8 per cent annually at global level; 40 per cent of mangrove forests in West Coast of India have been converted into farmlands and housing colonies over the last three decades.

Some of the mangrove species like Bruguiera cylindrica and Sonneratia acida are at the verge of extinction. Due to shrimp farming, about 35,000 ha of mangroves have been lost in India.

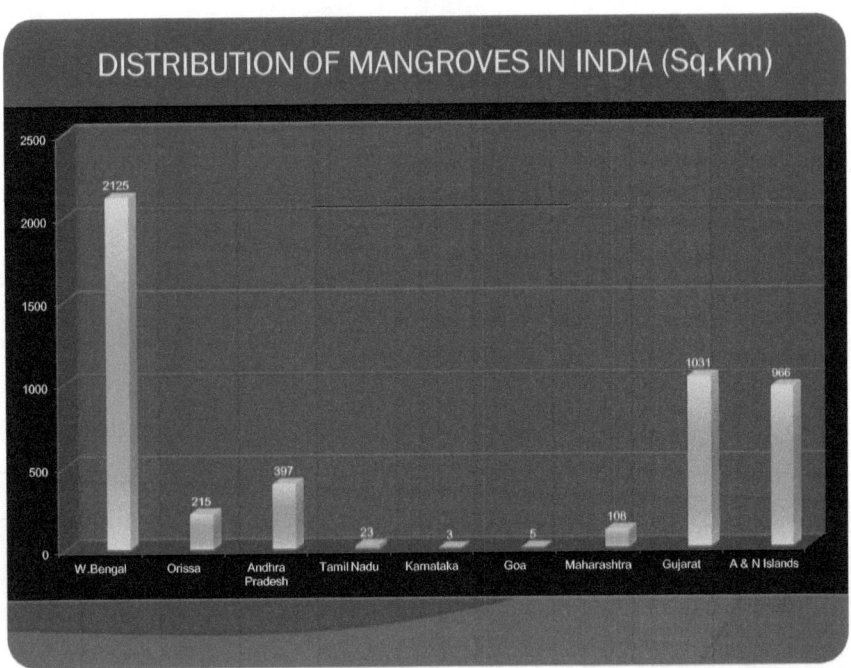

Though we have the Environmental (Protection) Act, the Supreme Court's order banning semi-intensive and intensive aquaculture in the coastal areas, and the Coastal Regulation Zone (CRZ) notification with strict enforcement of the legislative measures, scientific management practices are very much essential for conservation and sustainable management of the precious mangrove forests.

Suitable sites are to be identified for planting mangrove species. Mangrove nursery banks should be developed for propagation purposes. The economic and ecological values of the species are yet to be identified.

Environmental monitoring in the existing mangrove areas should be taken up systematically and periodically. Various threats to the mangrove resources and their root causes should be identified, and earnest measures should be taken to eliminate those causes.

The participation of the local community should be made compulsory for conservation and management. Floristic survey of mangroves along the coast is to be taken up to prepare biodiversity atlas for mangroves.

Potential areas are to be identified for implementing the management action plan for mangroves, especially in cyclone prone areas. Socioeconomic studies of the mangrove-dependent people need to be taken up to involve them in management of mangrove biodiversity.

Coastal industries and private owners need to be persuaded to actively participate in protecting and developing mangrove biodiversity. The forest department officials should be trained on taxonomy, biology and ecology of mangrove species.

Agro-forestry along the periphery of mangroves in the wastelands can be taken up for providing alternative fodder to the cattle of nearby villages. Alternative livelihood and income generation options like dairy farming, bee keeping, palm candy production, coconut leaf thatching, dry fish marketing, small provisional shops, vegetable shops, etc., can be taken up by the mangrove-dependent communities.

So far, none of the mangrove species has been included in the Red List of the International Union for Conservation of Nature (IUCN). A scientific study reported that 100 per cent of mangrove species, 92 per cent of mangrove associates, 60.8 per cent of algae, 23.8 per cent of invertebrates and 21.1 per cent of fish are under threat.

Periodical monitoring of the mangrove forest is very much necessary to assess the status. The impact of environmental and human interference on marine flora and fauna needs to be assessed.

The traditional rights of coastal communities to use the natural resources in their surrounding natural habitats for their livelihood should be recognised while formulating and implementing regulations and conservation measures on priority basis.

Therefore, while mangrove forests play a major role with more valuable ecological services, scientific management of the same is the need of the hour not only for the wellbeing of the mankind but also for coastal biodiversity.

Courtesy: Google

Pichavaram mangrove forests of Tamil Nadu

https://www.downtoearth.org.in/blog/wildlife-and-biodiversity/scientific-management-of-mangroves-is-need-of-the-hour-64007. April 19, 2019

28
ENACT A SPECIAL ACT FOR PROTECTING INDIA'S NATURAL PHARMACY

The country's medicinal plants, which are used in a number of alternative systems of medicine, are in danger of becoming extinct due to ignorance and over-exploitation.

The Ziziphus nummularia commonly known as the 'Jhar Beri' in Hindi and Urdu is used to treat nausea as well as scabies.

Photo: Wikimedia Commons

Ayurveda, Yoga, Unani, Siddha and Homoeopathy are generally called traditional medicine or alternative medicine in India. Though these systems of medicine have their own significance, they have some common specialities like zero side effects. They not only cure diseases but also root out the causes entirely, help cure the ailments through daily food habits, herbs and simple exercises and above all, are highly inexpensive.

Many of these alternative systems of medicines treat ailments with the help of nature's rich pharmacy that is available to them: medicinal plants and herbs.

India is not only known for its varied culture and traditions but also for its rich biological diversity. The large numbers of floral and faunal species bear testimony to its rich biological diversity. Of the many floral species, a large number are medicinal plants, which have been identified as our 'National Heritage'. They need to be conserved and managed sustainably.

The medicinal values of many plant species have been described in 'Rig Veda', about 4000 years ago. About 90% of the world's rural population depend on traditional herbal medicine for their primary health even today.

India is known as the emporium of medicinal plants due to the occurrence of several thousands of medicinal plants. Many of the tribal communities of the country like Irulas, Kanis, Kotas, Kurumbas, Malayaalis, Paliyas, Paniyas, Sholagas, Todas and others still rely on the naturally occurring medicinal plants for various diseases.

Mostof the medicinal plant species in India are found in forest areas. But there is no special act to protect these rare species except the Tamil Nadu Forest Act-1882 and the Wildlife (Protection) Act-1972.

Schedule VI of the Wildlife (Protection) Act-1972 covers only beddome's cycad, blue vanda, kuth, ladies slipper orchids, the pitcher plant and red vanda.

As the officials of the forest department are not given any special training about the medicinal plants, they are not even able to identify these plant species. When this is the case, even if any illegal removal of these plants takes place, they are not able to deal with it properly and effectively. This leads to indiscriminate destruction of the valuable plant species.

Hence effective and useful training of all the field staff and officers in identifying the medicinal plants and dealing with the occurrences involving the medicinal plants is very much necessary.

Due to indiscriminate exploitation, many of the medicinal plant species have attained the status of RET (Rare, Endangered and Threatened) species.

Therefore before they become extinct, it is the foremost duty of the state and the central governments to conserve these valuable medicinal plant species by enacting a special act and imparting training to the forest department officials not only for prosperity but also for posterity.

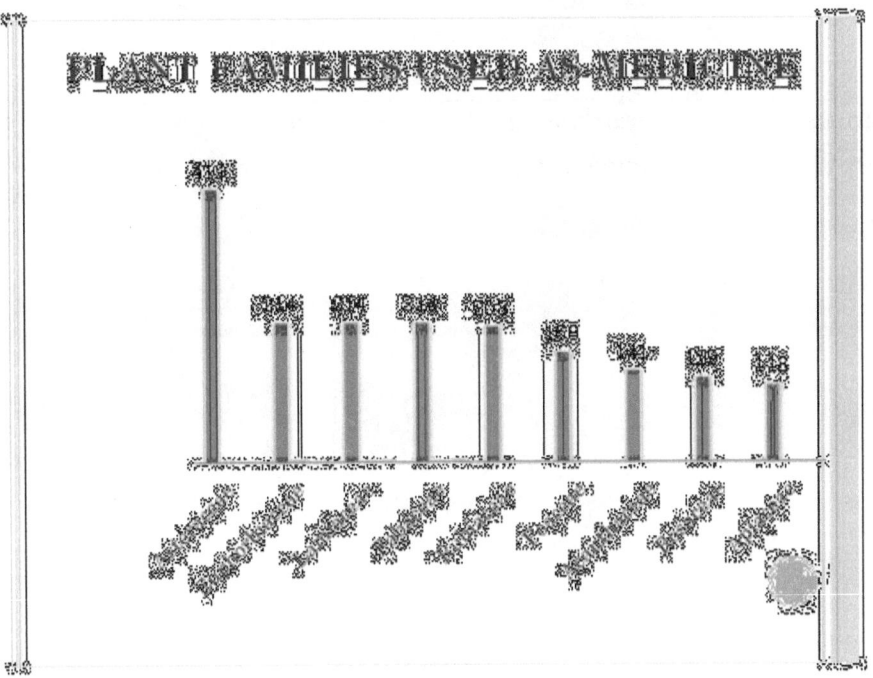

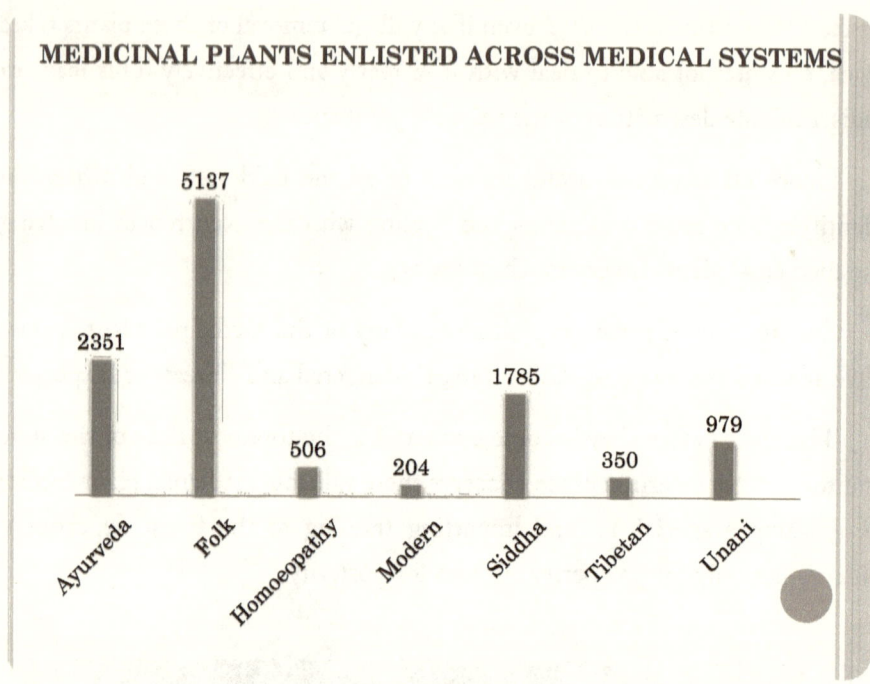

https://www.downtoearth.org.in/blog/wildlife-biodiversity/enact-a-special-act-for-protecting-india-s-natural-pharmacy-64768 May 28, 2019

29
CAUVERY: THE RIVER THAT THE TAMILS THOUGHT WOULD NEVER FAIL

The 'Ganga of South India' may be dying now, but was once a cherished waterbody in the lore of Tamil country

Courtesy: Google

Stanley Reservoir across Cauvery, also known as Mettur Dam across Cauvery

The Cauvery (also spelt as 'Kaveri') knowan as 'Ponni', in Tamil, is the fourth-largest river in South India. Originating in the Western Ghats at Talakaveri, in Karnataka's Kodagu district, it passes through Tamil Nadu. The river bisects the state into north and south and finally reaches the Bay of Bengal at Poompuhar, also known as Kaveripoompattinam in Tamil Nadu.

The Cauvery basin spreads over 81,155 square kilometres (sq km) in the states of Karnataka (34,273 sq km), Tamil Nadu (43,856 sq km) and Kerala (2,866 sq km) and the Union Territory of Puducherry (160 sq km).

The Cauvery's major tributaries, Kabini and Moyar join it before it reaches the Stanley Reservoir at Mettur in Salem district of Tamil Nadu. The river's total length, from source to mouth, is 802 kilometres.

From antiquity to the present era, the river has been the lifeline of the ancient kingdoms and cities of south India. Because of the river's bountiful nature, the Cauvery delta was considered to be the most fertile regions till recently.

Today, though, that is no longer the case: Karnataka and Tamil Nadu have been bitterly at odds with each other over sharing the river's waters due to the failure of the monsoon and erratic rainfall.

A cherished waterbody

Today, the Cauvery might be a minefield of controversies. However, it has been among the most loved, cherished and celebrated waterbodies, finding an important place in Tamil literature, right from the Sangam era, spanning from 5th BCE to 3rd century CE.

For instance, **Pattina Palai**, a Tamil poetic work belonging to the Sangam period, speaks highly of "Cauvery, the river that never fails, even if the sky does".

[Tamil]:வான் பொய்ப்பினும் தான் பொய்யா

மலைத்தலைய கடற்காவிரி

[Roman] "Vaan Poyppinum Thaan Poyya, Malaiththalaiya Kadar Cauvery".

Similar praise of the river is found in Puranaanooru and Porunaraatru Padai.

Purananooru, one of the eight books in the secular anthology of Tamil Sangam literature writes:

"The forceful water flow in the river Cauvery that can even uproot a tree, takes care of all the living beings of the world, as a woman feeds a child through her breast"

"புனிறு நீர் குழவிக்கு இலிற்று முலை போலச்

சுரந்த காவிரி மரங்கொல் மலிநீர்

மன்பதை புரக்கும்"-Poem No: 68.

"Puniru neer kuzhavikku ilitru mulai polach
Surantha kaaviri marangol malineer
Manpathai purakkum".

In the Kamba Ramayanam, when the great Tamil Poet Kambar compares Kosala kingdom of Sri Rama with the realms of the Cholas, he attributes the fertility of Chola kingdom to the perennial river Cauvery.

The Five Great Epics of Tamil Literature composed in Classical Tamil are Silappathikaram, Manimekalai, Seevaka Chinthamanai, Valayapathi and Kundalakesi.

In 'Silappadhikaram', composer Elangovadikal sings many songs in praise of the Cauvery and elaborately describes how it has made the Chola kingdom at the Cauvery delta fertile and flourishing.

Elangovadikal speaks about the greatness of the river Cauvery through Kovalan, the hero of the epic and Madhavi, the second heroine. While Kovalan sings about the Cauvery in only three poems, Madhavi greets the ruler of the country and the river Cauvery in three poems. Thus, controversy arises between them and Kovalan gets separated from Madhavi.

Says Madhavi:

"Long live this fertile country! Long live the Cauvery that nurtures the country like a mother".

"வாழி அவன்தன் வளநாடு
மகவாய் வளர்க்கும் தாயாகி
ஊழி உய்க்கும் பேருதவி
ஒழியாய் வாழி காவேரி"

And Kovalan's words are:

"As farmers sing;
As the sounds of sluice gates rise;
As new waters break open'
As people celebrate;
You walk! Long live Cauvery!"

"உழவர் ஓதை மதகோதை
உடை நீரோதை தண் பதங்கொள்
விழவரோதை சிறந்தார்ப்ப
நடந்தாய் வாழி காவேரி"

In Manimekalai, another famous epic, the poet Saaththanaar describes the Cauvery with due regard and reverence.

"As per the request of Kanthaman, the then Chola King, the great sage Agasthiyar (Agastya) poured holy water from his kamandal and the water started flowing in the name of the river Cauvery".

"Kanjavetkaiyin Kanthaman venda
Amara munivan Agaththiyan thanaathu
Karagam kavizhththa Cauveryp pavai"-

"கஞ்சவேட்கையின் காந்தமன் வேண்ட
அமர முனிவன் அகத்தியன் தனாது
கரகம் கவிழ்த்த காவிரிப் பாவை"

On seeing the river, the Goddess Sambapathy welcomes Cauvery thus:

"Welcome the Ganges from the blissful sky that fulfilled the desires of the king Kanthaman".

"Aanu visumbin aagaaya gangai!
Venavaath theerththa vilakke! Vaa!"-

"ஆணு விசும்பின் ஆகாய கங்கை!
வேணவாத் தீர்த்த விளக்கே! வருக!"

The poet describes the Cauvery as "The river that nourishes the land with perennial water, even if the zodiac signs show impending severe summer". He praises the river Cauvery as the family deity of the Chola Kingdom because of its boundless benevolence through its copious flow of water".

"Paadal saal sirappin Bharathaththu ongiya
Kodaach sengol chozhartham kulakkodi
Kolnilai thirinthu kodai needinum
Than nilai thiriyaath than thamizhp paavai"-

"பாடல் சால் சிறப்பின்பரதத்து ஓங்கிய
கோடாச் செங்கோல் சோழர்தம் குலக்கொடி
கோள்நிலை திரிந்து கோடை நீடினும்
தன் நிலை திரியாத் தண் தமிழ்ப் பாவை"

Chola King, Karikalan (Karikala Cholan) built a stone dam across Cauvery 2000 years ago based on the representations received from peasants to increase the extent of paddy cultivation.

This dam was called the 'Grand Anicut' by Sir Arthur Thomas Cotton, the famous British Engineer and served as a model for constructing a dam across the river Kollidam near Mukkombu during the 19th century. After the construction of the dam and other canals, about 1.2 million acres of land were brought under cultivation.

Many temples like the Sri Renganatha Swamy temple at Srirangam (the island in the middle of the Cauvery in Tiruchirappalli), Pragatheeswara temple at Thanjavur and Sri Nataraja temple at Chidambaram had been built along the course of the Cauvery as well as in its delta.

As the Chola Empire was very prosperous due to Cauvery, the lifeline of the kingdom, the then rulers were able to extend their ruling up to Southeast Asia.

Memories and the death of a river

During our training period at the erstwhile 'Southern Forest Rangers College', Coimbatore, sometime in 1975, I happened to stay with others in tents in Makut (near Brahmagiri) forests of Karnataka, from where the Cauvery originates. The site was located just on the bank of the river.

The forests were of moist deciduous type and water flow in the river was said to be perennial. But, of late due to the development of roads, railway tracks, electricity pylons, etc., right across the rain forests, it has lost its charm and greenery. Moreover, the rainfall levels have come down drastically, resulting in hardly any water in the river.

In the past, every year on June 12, water used to be released from the Mettur dam, enabling farmers in the Cauvery delta to go ahead with the farming activities. Now it is only a memory.

Water was shared between the states of Karnataka and Tamil Nadu according to agreements signed in 1892 and 1924 between the Madras Presidency and the Wodeyar Kingdom of Mysore. About 44,000 sq km of the river basin is in Tamil Nadu and 32,000 sq km in Karnataka.

Since the water sharing has been done according to a British era agreement, Karnataka does not agree with the sharing of drinking water with Tamil Nadu. Karnataka also wants the agreement to be renewed according to the present day rainfall patterns.

The Supreme Court of India, in its verdict dated February 16, 2018, had ruled that Karnataka would get 284.75 thousand million cubic feet (tmcft), Tamil Nadu 404.25 tmcft, Kerala-30 tmcft and Puducherry-7 tmcft.

The apex court also directed the Centre to notify the Cauvery Management Scheme. The central Government notified the 'Cauvery Water Management Scheme' on June 1, 2018, constituting the 'Cauvery Water Management Authority' and the 'Cauvery Water Regulation Committee'.

The Cauvery Water Management Authority (CWMA) ordered Karnataka on May 28, 2019 to release 9.19 tmcft of water to the lower riparian states for June. The CWMA meeting was attended by the representatives of the Centre and the riparian states, Tamil Nadu, Kerala, Karnataka and Puducherry. As several parts of Tamil Nadu are reeling under acute water shortage, the Centre had already issued a drought advisory, warning the depleting water levels in the reservoirs.

Karnataka has to ensure flow of about 3 tmcft to Tamil Nadu by June 10. But, it has only about 14.43 tmcft of water in the four major reservoirs of Cauvery basin, Krishna Raja Sagar, Kabini, Hemavathi and Harangi. They need 4.84 tmcft of water to meet out the drinking water requirements to 47 towns, including Bengaluru, Mysuru, Mandya and 625 villages in the basin area.

Karnataka has already issued directions not to use any water for irrigation. Many talukas in the basin have been declared as drought-hit. As such, Karnataka is not able to release any water to Tamil Nadu. It states that only if the catchment areas receive good rainfall, will Karnataka think of releasing the due share to other states.

In the present situation, only a good monsoon with copious amount of rainfall can help all the concerned parties. It is high time that everyone realises the responsibility of using the water judiciously, stop polluting the water and taking all possible efforts to increase the ground water table.

Effective water management is the need of the hour. May be then, the Cauvery will be bountiful again one day.

V.Sundararaju is President, the Society for Conservation of Nature, Trichy, Tamil Nadu, and Consultant with the Society for Social Forestry Research & Development, Tamil Nadu.

https://www.downtoearth.org.in/blog/water/cauvery-the-river-that-the-tamils-thought-would-never-fail-64973 June7, 2019

30

CHENNAI WATER CRISIS: SOME SUGGESTIONS TO AVERT A REDUX

Here are some suggestions to see to it that another incident like Chennai going dry does not happen in the future

The scenes beaming in from Chennai, the capital of Tamil Nadu and the sixth-largest city in India earlier this summer were disturbing. Bone-dry reservoirs, people lining up to fill vessels with meagre amounts of water and even some cases of violence, shocked India and the world.

Conserve traditional water bodies

Due to an ever-increasing population, wetlands, water bodies and even rivers have been encroached upon in each and every city and village of India.

Out of the two dozen water bodies and wetlands once existed in Chennai, only nine exist currently. Similarly most of the fresh water marshes and lakes in the Gangetic flood plain, the biggest river plain of India, have been lost.

Consequently the capacity of water storage has come down drastically.

Stern action has to be taken to evict the encroachments on water bodies. Sincere and speedy action is the need of the hour to deepen these water bodies.

Special efforts should be made for removal of the weeds like Prosopis juliflora and water hyacinth from the water bodies.

Practice rain water harvesting

Rain Water Harvesting (RWH) was given importance during 2003 by the then Tamil Nadu Government, according to advice given by veteran agricultural scientist MS Swaminathan.

Building approval was granted only for those upcoming buildings which had RWH structures. This was made compulsory for existing buildings also.

But, most of such structures were equipped with RWH only in name, without serving any useful purpose. In recent years, when a study was undertaken, many such RWH structures built in government offices were found to be not functioning.

According to a study made by the Central Water Commission, the annual requirement of 3,000 billion cubic metres of water can be managed from the annual rainfall of 4,000 billion cubic metres, if the rainwater is harvested properly. But, we are harvesting only eight percent of the rainwater, which is the lowest in the world.

Practice drip irrigation

It is estimated that 80 percent of the fresh water in India is used for agriculture. The drip irrigation method has been found to be more effective for raising agricultural crops successfully. Besides helping to save 70 percent of water, 30-40 percent of electricity is also saved through this system. Above all, income to the agricultural community is also increased doubly through increased crop productivity of about 30-90 percent. Unfortunately, till 2016-17 drip irrigation was used in only four percent of the total irrigated area.

Recycle and reuse wastewater

Since recycling and reuse of the domestic wastewater is not done properly in India, around 80 percent of the water is allowed through sewage to join the water bodies like streams and rivers.

Rivers including the Ganga, Yamuna and Cauvery, while providing the drinking water supply, are polluted by effluents, sewage and plastic waste released by the factories and households.

For instance, the Cauvery provides drinking water to about 14 districts in Tamil Nadu. But ironically the very same Cauvery is polluted through various tributaries like Amaravathy, Noyyal and Kudamurutti.

Only through stringent action initiated by the judiciary, this kind of pollution can be brought under control.

Teach water harvesting to farmers and the others

Water harvesting methods are to be taught to the farmers and other common public. Farm ponds are to be created at least to an extent of 10 percent of the landholdings.

Practice judicious water management

Judicious water management is of great help in conserving ground water, even during drought. The Kodikulam village near Madurai in Tamil Nadu sets an example in water management.

Though the depth of the well is only 18 feet, it never dries up. If any bore well is sunk, water is found available at a depth of 50 feet. The villagers follow certain discipline in drawing water from the well.

Water is drawn daily before 9.30 am in the morning and after 4.30 pm in the evening. The farmers do not sink any bore well for irrigation. Agriculture is carried out only with the support of the water released from the Periyar dam.

Another interesting example is Easwari Nagar near Pallavaram in Chennai. The villagers fulfil their water requirement only from the hundred-year-old well dug out during the British period.

Though the well was abandoned after independence, it was reclaimed during 1975, a time of water scarcity. From then onwards, the well has been taken care of by the residents of the village. Water is drawn from 7 to 8 am, 11 am to 1 pm and from 6 to 8 pm. Only 3 pots of water are allowed for each family.

We should create awareness among the people about judicious water management following the experiences of Cape Town, South Africa through the concept like Day Zero.

Day Zero refers to the day when a place is likely to have no drinking water of its own. According to a report released by NITI Aayog last year, 21 cities in India may run out of groundwater by 2020. Therefore it is high time that the government should rise up to the occasion to meet the water requirement of the people before it leads to economic losses and social unrest.

Courtesy: Google

Acute water scarcity as experienced in Virudhunagar District of Tamil Nadu

https://www.downtoearth.org.in/blog/water/chennai-water-crisis-some-suggestions-to-avert-a-redux-65897 Monday 29 July 2019

31
WHY THERE IS AN ALGAL BLOOM IN GULF OF MANNAR AND HOW IT AFFECTS LIVELIHOOD

A recent algal bloom in the Gulf of Mannar, caused allegedly due to pollution from Sri Lanka, highlights how fragile our coral reefs are.

Tamil Nadu is blessed with the second-longest coastline in India -1,076 kilometres stretching from Pazhaverkadu in Thiruvallur district to Ezhudesam in Kanyakumari district.

Sixty percent of the state's population live within hundred kilo metres of the coast. The rivers which originate from the Western and Eastern Ghats join the sea in the state.

About 26 towns and 2390 villages are located along the coast. About a million fishers in Tamil Nadu, who live in 608 fishing villages, depend on fisheries for their livelihood. The livelihood of the fishers is threatened due to depletion of natural resources owing to habitat destruction, pollution, and other factors.

In September 2019, the fishing community in Ramanathapuram, Tamil Nadu, was shocked as the sea water turned green and fish died in the thousands.

Scientists from the Central Marine Fisheries Research Institute (CMFRI) in Kochi visited the spot, carried out scientific study and came to the conclusion that the incidents had taken place due to the sudden blooming of Noctiluca, a type of marine microalgae.

The cause of the fish deaths was assessed to be oxygen depletion caused by the sudden blooming of microalgae. The fishers were advised not to worry as the algae would dissipate when there was a downpour and strong currents in the sea.

Scientists from the National Centre for Coastal Research (NCCR), Mandapam have said that the Noctiluca scintillans algae bloomed suddenly due to the discharge of ballast water, subsequently causing the fish to die-off and the sea water turning green.

Blooming of the algae took place between the Gulf of Mannar and Mannar areas off the Sri Lankan Coast. It could be possible that the discharge of the ballast water was from the Lankan coast.

According to forest department sources, it has confirmed that while coral reefs were found dead at Shingle Island near Rameswaram after blooming of the algae, no such casualty was noticed in Kurusadi Island, south of Pamban.

The Reef Research Team of the Suganthi Devadasan Marine Research Institute (SDMRI) has reaffirmed that algal blooming had killed about 180 coral reef colonies in Shingle Island.

Of corals and algae

The Ramanathapuram incident is just one example of how anthropogenic activity can cause great damage to the fragile marine ecosystem.

Algal bloom caused by nutrient pollution (nitrogen and phosphorous) can cause great harm to aquatic life due to the toxic content which they possess. While some algae can make animals sick, other creatures can die off in large quantities and deplete oxygen during the process of decomposition. Climate change also plays a major role in algal blooming.

When the algal bloom blocks the sunlight from reaching the algae within the coral, they cannot photosynthesize and create food for the corals. Besides, the depletion of oxygen due to algal decay also can have adverse impact on the coral.

As the coral reefs are damaged, they are not able to provide food and shelter to fish and other aquatic life. As a result, the livelihood of millions of people who depend upon the marine resources is jeopardized.

The harmful effects of algal blooming on coral reefs may be devastating and emergent attention should be paid to minimise our contributions to climate change and nutrient pollution in order to give the coral reefs a new lease of life.

The anthropogenic factors which are responsible for destruction of the marine environment are sewage, the plastic menace, sedimentation, industrial pollution, thermal pollution, salt pans, oil pollution, destructive fishing practices such as over-fishing, dynamite fishing, poison killing, trap fishing, bottom trawling, coral mining, etc.

The following environmental laws have been implemented for conserving the marine environment: The Indian Fisheries Act-1897, The Wildlife Protection Act-1972, The Water (Prevention and Control of Pollution) Act-1974, The Environment (Protection) Act-1986, The Coast Guard Act-1950 and Coastal Regulation Zone (CRZ) Notification-1991.

The Government of India has signed and ratified several international conventions relating to oceans and related activities namely the United Nations Convention on the Law of the Sea (UNCLOS) in 1982, International Convention for the Prevention of Pollution from Ships (MARPOL), 1973-1978, the London Dumping Convention in 1972, the Convention on Civil Liability for Oil Pollution Damage (CLC) in 1969 and the Convention on Biological Diversity in 1992.

MARPOL strives to protect the marine environment through elimination of discharges of oil and other harmful substances. MARPOL has recognized GoM as 'Special Area' where discharges are especially restricted.

The issue of algal blooming due to release of ballast water from the ships off the coast of Sri Lanka is to be taken up with the government of Sri Lanka to prevent further damage to the marine environment on Tamil Nadu side.

The pollution control board, the forest department and other connected agencies are to be geared up to monitor, regulate and control the harmful algal blooming caused by the release of pollutants from the salt manufacturing industries, aquaculture firms, chemical industries, etc., by implementing the related laws effectively.

Courtesy: Google

Coral Reefs killed by microalgae

https://www.downtoearth.org.in/blog/pollution/why-there-is-an-algal-bloom-in-gulf-of-mannar-and-how-it-affects-livelihood-67461 Monday 28 October 2019

32
WHY SOUTH INDIA NEEDS THE SHOLA FORESTS OF THE NILGIRIS

The forests and grasslands act as water towers and influence the fortunes of farmers in the Cauvery delta.

The Shola forests of South India derive their name from the Tamil word solai which means a 'tropical rain forest'. Classified as 'Southern Montane Wet Temperate Forest' by experts Harry George Champion and SK Seth the Sholas are found in the upper reaches of the Nilgiris, Anamalais, Palni hills, Kalakkadu, Mundanthurai and Kanyakumari in the states of Tamil Nadu and Kerala.

These forests are found sheltered in valleys with sufficient moisture and proper drainage, at an altitude of more than 1,500 metres. The upper reaches are covered with grasslands, known as Shola grasslands.

The vegetation that grows in Shola forests is evergreen. The trees are stunted and have many branches. Their rounded and dense canopies appear in different colours.

Generally the leaves are small in size and leathery. Red coloured young leaves turning into different colours on maturity is a prominent characteristic of the Shola forests. Epiphytes like lichens, ferns and bryophytes are found grown on the trees.

The occurrence of Himalayan plants like Rhododendron is a mystery. Paleobotanist Vishnu Mitter suggested that these are remnants of the vegetation driven to South during the Quaternary Ice Age, about 2.6 million years ago, with subsequent changes in the tropics of South India.

Sholas play a major role in conserving water supply of the Nilgiris' streams. In the Madras District Gazetteers-The Nilgiris (1908), W. Francis says, "The Sholas of the plateau are not of any great importance from a commercial point of view, as the trees are slow-growing varieties which produce timber of little or no value and probably take at least a century to mature. But they add greatly to the beauty of the country and are of immense use in protecting source of water supply".

Sholas thus act as the 'Overhead Water Tanks'.

The rolling grasslands found on top of the Western Ghats, enhance the beauty of the region. Generally Shola forests and grasslands are found in the ratio of 1: 5.

Pastoral communities, who settled down in the grasslands centuries ago, periodically burn grass. This has checked the advance of the Shola forests. As tree species of the montane evergreen forests are flammable, regeneration of any Shola tree species is completely prevented except for Rhododendron nilagiricum, the only Shola tree which can tolerate fire.

The rain received from the Southwest and Northeast monsoons is harvested by the Shola forest-grassland ecosystem, leading to the formation of the Bhavani river that finally drains into the Cauvery. Thus, the Shola forest-grassland ecosystem of the Nilgiris, also supports the prosperity of Cauvery delta farmers.

Trouble for Sholas

Unfortunately, the Sholas have begun to gradually shrink due to the introduction of alien plant species and annual fire occurrences.

Alien species like Sticky Snakeroot, Gorse and Scotch Broom introduced during the British rule, have encroached upon the grasslands.

During 1840, tree species such as Acacia and Eucalyptus were introduced from Australia. Afterwards, between 1886 and 1891, Pine and Cypress were introduced again from Australia. As the alien species grew, the forests and grasslands gradually became degraded and shrunk.

In addition, unscientific agricultural practices like growing tea on the slopes, cattle grazing and fuel wood collection have become serious causes for degradation. Unregulated tourism has created concrete jungles, traffic congestion and caused the generation of garbage.

Land-use studies undertaken on the Nilgiri Biosphere Reserve between 1849 and 1992 show the extent of the damage. During 1849, the extent of Shola forests was 8,600 hectares (ha), grasslands 29,875 ha and agriculture was 10,875 ha. No wattle or eucalyptus was planted in the area at that time.

During 1992, it was found that the extent of Sholas was 4,225 ha, grasslands 4,700 ha, agriculture 12,400 ha, tea plantations 11,475 ha, wattle plantations 9,775 ha and eucalyptus plantations was 5,150 ha.

The comparison of the results of 1849 and 1992 studies shows that cultivation of tea, wattle and eucalyptus has reduced the Shola forest-grassland ecosystem to a great extent.

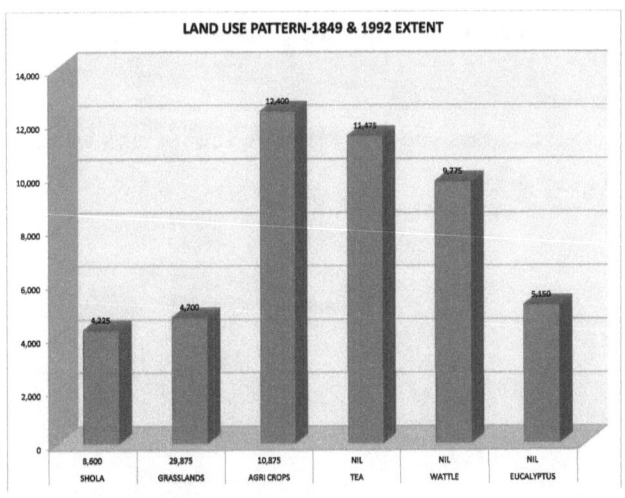

Measures taken

After realising the seriousness of the situation, the government banned the planting of wattle and eucalyptus completely in 1987. Ecological restoration and biodiversity conservation were given importance.

Under the Hill Area Development Programme (HADP) since mid-1980s, seedlings have been planted in degraded patches and protected with chain-link fences to restore the forests.

Special Shola Forest protection Committees were formed involving teachers, nature lovers, ecologists, environmentalists, students and villagers in the Nilgiris. They were motivated to remove plastic garbage from the nearby forests, protect Shola trees, remove alien species and learn the importance of the Sholas.

The forest department started supplying LPG to villagers who lived near the Sholas as they depended upon the forests for their fuel wood needs. This helped the forests a great deal as the entry of the people in them was stopped.

Presently, the Tamil Nadu forest department focuses on eradicating wattle, providing fencing and planting Shola seedlings in degraded Shola forests.

The Shola-grassland ecosystem, which acts as the Nilgiris' overhead water tank and the water source for the Cauvery Delta, can be saved with the involvement and cooperation of the common public, students and nature enthusiasts.

If the forests and grasslands are restored, the region's water problems will be solved to a great extent.

Photo by Author

Shola forests near Avalanche in the Nilgiris

https://www.downtoearth.org.in/blog/forests/why-south-india-needs-the-shola-forests-of-the-nilgiris-68948 24 January 2020

33

INDIA NEEDS TO PROTECT ITS WETLAND FLORA

Wetland flora helps maintain ecological balance and need to be conserved.

However, only 68 wetlands are protected in India so far.

Wetlands in India are facing an ecological imbalance. Uncontrolled siltation; discharge of waste water and industrial effluents; denudation of aquatic vegetation; aquaculture; construction of dykes, dams and seawalls; discharge of pesticides, herbicides, etc; and filling for solid waste disposal-these are some of the major threats to wetlands.

Wetlands are a distinct ecosystem flooded by water. Its aquatic vegetation makes it different from other aquatic bodies.

Aquatic system flora helps maintain ecological balance by interacting with their environment in numerous ways, and therefore, their management is crucial. So far, only 68 wetlands are protected in India. Thousands of other wetlands that are biologically and economically Olympian have no legal status.

Flora vegetation

Aquatic (fresh water) vegetation-which grows around streams, rivers, lakes, etc-has more ecological significance than sea-water vegetation. The former maintains water quality by filtering out nutrients and sediments.

They also play a role in the food web. Seeds or tubers found in fresh-water ecosystem are consumed by waterfowl species. Many plants enter the food chain as detritus, which are small plant particles formed after decomposition of plants and are subsequently consumed by invertebrates.

Fresh-water vegetation, thereby serves as a breeding ground for aquatic and terrestrial fauna. It provides nesting areas for migratory birds. They help prevent erosion and stabilise soil.

Wetlands flora can broadly be classified into submerged water plants, floating water plants, emergent water plants and riparian water plants.

Submerged water plants are 100 per cent under water, and provide a food source for native fauna and habitat for invertebrates. They also possess filtration capabilities.

Floating water plants are found in slow-moving water and have small roots. They are a source of food for avian species. Emergent water plants grow above water with their roots submerged in marsh localities. Surrounding trees and shrubs found along the edges of wetlands or other water bodies are called riparian water plants.

The marine ecosystem comprises deltas, coral reefs, mangrove forest, lagoons, sea grass beds, etc. Seaweeds and seagrass are the major plants found in sea-water.

While seaweeds are primitive, marine non-flowering plants (algae) without roots, seagrass are flowering plants that grow submerged in shallow coastal waters and estuaries.

Urchins and fishes feed on seaweed, which also provide shelter to fishes, invertebrates and mammals. Large seaweeds can form dense underwater forests, called 'Kelp forests' and act as underwater nurseries for many marine animals, such as snails and sea urchins.

Seagrass, on the other hand, require sunlight for photosynthesis. They have tiny flowers and strap-like leaves. Less than 60 species globally, they help maintain water clarity by trapping fine sediments and particles with their leaves and help prevent soil erosion. They provide habitat for many fishes and invertebrates. Seagrass and other organisms that grow on them are a food source for many marine animals.

Mangroves, which are part of marine ecosystem, serve as breeding, feeding and hiding place for many fishes, crabs, oysters, prawns, etc. Apart from protecting the coastline from erosion, they control the floods also. Halophytes act as salt extractors and reduce soil salinity.

The road ahead

To maintain the ecological equilibrium, it is crucial to maintain wetlands. The problem, however, is that India's wetlands have not been delineated properly so far.

The Ministry of Environment, Forests and Climate Change is primarily responsible for the management of wetlands. Effective coordination between the different ministries such as energy, industry, fisheries, revenue, agriculture, transport and water resources is essential for protecting the ecosystems.

Only the wetlands that come under the Protected Area Network have Management Plans; but others do not. Their active monitoring over a period of time is essential. Comprehensive inventory of all the wetlands involving the flora, fauna and biodiversity, along with direct and indirect values, should be prepared. There is no special legislation to protect these ecosystems.

An environmental impact assessment plan needs to be prepared highlighting threats to these ecosystems and formulating corrective measures. As wetlands are common property with multi-purpose utility, their protection and management need to be a common responsibility.

An appropriate forum to resolve existing issues needs to be set up. All relevant ministries need to allocate sufficient funds for conserving these ecosystems. Awareness among the general public, educational and corporate institutions must be encouraged to achieve sustainable success in protection of these wetlands.

Microwave remote sensing tools play an important role in applications relating to wetland monitoring and vegetation assessment. Microwave sensors are highly sensitive to moisture content and textural properties of vegetative cover. They can be used to discriminate between grasses, aquatic vegetation, forest and crop cover.

Encroachments can be identified. Unmanned Aerial Vehicle (UAV) technology can be tried for monitoring the wetlands. Remote sensing data in combination with GIS methods are effective tools that have been used to delineate the open water habitat with aquatic vegetation in Keoladeo Ghana National Park in Bharatpur.

It is high time that earnest efforts are taken on scientific basis for better management of flora of wetlands.

Courtesy: Google

Flora of Wetland

https://www.downtoearth.org.in/blog/wildlife-and-biodiversity/india-needs-to-protect-its-wetland-flora-69469 Wednesday 26 February 2020

34
A UNIQUE INSTITUTION FOR A BELOVED INSECT

The Tropical Butterfly Conservatory in Tiruchirappalli is a boon to the people of India.

While there are about 1300 bird species and 370 mammal species in the Indian subcontinent, the diversity of Indian butterflies is assessed to be about 1501. The major threats to butterfly diversity are destruction, degradation and fragmentation of their habitats, grazing, fires and application of pesticides and weedicides in agricultural and urban ecosystems.

As butterflies form an important part of nature's food web, it is very much essential to protect the species for ecological balance. With this in mind, the Tropical Butterfly Conservatory Tiruchirappalli (TBCT) has been developed in Tamil Nadu's Tiruchirappalli to create awareness among the public about the importance of butterfly and its ecology.

The Conservatory is located in the Upper Anaicut Reserve Forest, sandwiched between Cauvery and Kollidam rivers in Tiruchirappalli. It is about 7 kilometres from Melur and is spread over 27 acres. This is considered to be Asia's largest butterfly park.

It was inaugurated during November, 2015 at Tiruchirappalli with the objective of propagating the importance of butterflies and conserving the biodiversity of the district through environmental education. The park has an

Outdoor Conservatory, Indoor Conservatory, a 'Nakshatra Vanam' and a 'Rasi Vanam' in addition to a breeding lab for non-scheduled species, an open air theatre, an amphitheatre, an interpretation centre, plant nursery, shade houses, ponds, water fountains, models, an eco-shop and a children's infotainment park.

So far, about 109 butterfly species have been observed here. Eggs of non-scheduled butterfly species are collected and bred in captivity in the in-house incubation laboratory by keeping them in ventilated plastic containers with the leaves of host plants as feed.

After attaining the transformation of larva (caterpillar) and pupa (transition), the adult butterfly finally comes out with gorgeous colours and at this stage they are released into the natural habitat.

Every now and then, non-scheduled butterfly species are bred and released by the park authorities into their natural surroundings. A Junior Research Fellow (JRF) with a background in entomology is engaged for breeding the non-scheduled species and for monitoring the butterflies.

The JRF serves as a guide for the visiting school and college students as well. Recently I visited the park during February, 2019 and had the rare opportunity of releasing newborn adult non-scheduled butterfly species that had been bred in captivity.

The water fountains erected at salient points of the park help to develop the cooling humidity suiting the requirements of the butterfly species during daytime. Butterflies are found to be active generally from 9 am to 11 am in the forenoon and from 3 pm to 6 pm in the afternoon collecting nectar and suitable nectar plants have been planted sufficiently to meet the insects' requirements.

During hot hours of the day, the insects take rest in shaded areas and plants suiting their roosting needs have been planted in plenty at the conservatory. Besides this, larval host plants also have been carefully selected and planted in required numbers in this park and the visit of increased numbers of butterflies is testimony to this.

The park has 298 plant species. A JRF with a background in horticulture is engaged for maintenance of the plant species.

Out of the 1501 butterfly species identified so far in India, 109 species have been observed in this park. As butterflies are known for their intrinsic, aesthetic, educational, scientific, ecological, health and economic values, they are considered to be the most universally loved and inspirational of all creatures.

Since butterflies play a key role in pollination of the plants species, the global food chain depends on their well-being.

The park has been created in such a way that it provides congenial atmosphere for the butterflies for breeding, procreating and completing their life cycle in the natural surroundings. The park is managed by the District Forest Officer, Tiruchy forest division. According to the forest officials of the park, the number of butterflies that visit here has also increased in recent years.

The handbook published on the Tropical Butterfly Conservatory Tiruchirappalli by the District Forest Officer is of great use to the visitors.

Nakshatra Vanam and Rasi Vanam have been developed with 27 trees corresponding to 27 stars and 12 trees corresponding to 12 zodiac signs of Indian astrology. These gardens have been created with the idea of motivating the people to plant and nurture trees associated to their stars and zodiac signs.

While the seeds are collected for further propagation, the dry leaves and the bark are composted and used as manure for the plants. The amphitheatre in the park is utilised for screening films on the butterfly's life cycle and the insect's key role in maintaining the ecological equilibrium.

About 2.5 lakh people visit the park annually according to park authorities. A one-day certificate course has also been conducted fortnightly from 2017 onwards on basic 'Lepidopterology' (Study of butterflies).

The visitors are advised to move inside the park either in the mid-morning or in the mid-afternoon to enjoy watching the butterflies at close quarters. They are instructed to wear dull coloured clothes in order to avoid any disturbance to the insects.

In a nutshell, the Tropical Butterfly Conservatory Tiruchirappalli can be described as a boon not only to the people of the district but also for the entire country.

Courtesy: Google

Butterfly sucking nectar from flower

https://www.downtoearth.org.in/blog/wildlife-and-biodiversity/a-unique-institution-for-a-beloved-insect-70207 Friday 03 April, 2020

35
KEEPING THE 'GREEN HILLS' TRULY GREEN

An eco-tourism model allows tourists to explore the wilderness of Tamil Nadu's Pachamalai hills and locals to have a stake in protecting them

The 'Pachamalai' hill range located in Tiruchirappalli district of Tamil Nadu means 'Green Hills' in English. These are unexplored areas of the Eastern Ghats, covered with dry evergreen and dry deciduous forests.

The concept of eco-tourism introduced in recent years by the forest department, involving the local forest stake-holders is advantageous not only for the residents but also for the biodiversity of the entire hill range.

The deteriorating pristine environment of Pachamalai hills due to unregulated tourism can be protected only if the local people realise the importance of the clean environment and cooperate with the forest department.

The range

The ecological significance of the Pachamalai hills is huge. With the average annual rainfall of 800-900 millimetres, mostly from the northeast monsoon, the forests serve as a catchment area for about 30 lakes situated at the foothills.

Numerous streams originate from the forests of the hill range. Their water irrigates thousands of hectares of land at the foot hills and the nearby plains.

The well-being of these water systems is closely related to the prosperity of the farmers of the district as the economy of the district depends on agriculture.

The rainwater due to precipitation in the reserve forests is collected by umpteen numbers of streams. The water sources can be maintained properly only if the reserve forests are protected well.

But, wildlife numbers in the area have reduced in the past due to biotic pressures like conversion of private forests into coffee plantations and orchards and encroachments by private estates.

Many of the animal species have been driven to the verge of extinction due to hunting, poaching and habitat loss in the past.

The rich biota and the fragile ecosystem of the hill range can be saved from the ruthless hands of wealthy and political mafia only by the timely intervention of the forest department.

Eco-tourism in the Green Hills

A strategy has now been devised to protect the area through the support and cooperation of the local tribes. They are engaged as guides for trekking and other eco-tourism allied activities.

The fabulous cuisine of the tribes, their amazing lifestyle and interesting culture are exposed to the visitors. Tourists, who stay here for a night, can enjoy the awesome campfire with the dance, song and music of the local tribes.

At the foothills of Pachamalai, forest rest houses commanding panoramic view of the hill range are available for the accommodation of the tourists.

Similarly, in Top Sengattupatti village located at the top of the hill, the forest department has developed 'Tree Top Huts', a dormitory and a British Era heritage rest house under eco-tourism.

The Tree Top Huts have excellent model suites for the comfortable stay of tourists. The dormitory can accommodate large groups of tourists who undertake trekking.

The British Era rest house, built amidst mahogany groves, has got its own heritage value. Even during hot summer, the lofty and gigantic mahogany trees provide a pleasant atmosphere and shelter from the heat and glare of sunlight through dense shade.

The heritage rest house consists of two suites has been developed with all facilities for convenient stay of the visitors.

Selected trekking routes through verdant and lush green reserve forest make one enjoy the natural surroundings. Trekking tours are conducted by the forest department in selected routes.

One forest guard and one forest watcher along with two tribal trekking guides accompany the tourists during the trekking. En route, important trees have been identified and provided with information boards containing the details of their scientific names, local names, families and uses, so as to enable the visitors to know about such trees.

Spotted deer, mouse deer, barking deer, sloth bear, slender loris, jungle cat, Indian giant squirrel, toddy cat, common monkey, common mongoose, blacknaped hare and porcupine are some of the commonly sighted animals while trekking.

Birds like Junglefowl, barbets, flycatchers, king crows, babblers, mynas, munias, flowerpeckers, sunbirds, parakeets, woodpeckers, rosy pastor, warblers, Indian pitta and kingfishers are some of the interesting birds that can be seen in the forest.

Eco-tourism in the Pachamalai Hills gives an opportunity to tourists to savour natural surroundings and experience traditional tribal life, while providing employment to locals and keeping the environment safe at the same time.

Photo by Author

Forest road with tall trees on either side near British Era heritage rest house at Pachamalai

https://www.downtoearth.org.in/blog.wildlife-and-biodiversity/keeping-the-green-hills-truly-green-71070

Tuesday 12 May 2020

36
TAMIL NADU MUST SUSTAINABLY MANAGE ITS RATTANS

The plants have seen huge commercial exploitation in the state and need restoration to safeguard their survival.

Rattans are climbing spiny palms with about 600 species belonging to 13 genera that are distributed all over the world. They are found in a number of countries and regions ranging from Australia to West Africa.

Rattans are an integral part of the tropical forest ecosystem. The numerous pinnate leaves (leaves growing on both sides of a common axis), with a length of about two metres, intercept the splash effect of rains and improve the water-holding capacity of the soil. They play a key role in enriching the soil through their leaf litter.

The fruits are relished by monkeys, squirrels and birds.

Importance to humans

Besides their ecological role, rattans have a wide variety of use in human societies. In certain parts of the world, the fruits and rootstalks are eaten by people. Some of the Rattan species have medicinal properties.

Their most important use is in the handicraft industry. They have been believed to have been used by humanity since the fifth century BC for making household articles, furniture, tool handles and bridge construction as well as sports goods like javelins, cricket bats and hockey sticks in modern times.

Rattans are important in handicraft and furniture-making because of its unique characteristics such as strength, durability, looks and bending ability. As they possess high value and social and economic importance, they are regarded as 'Green Gold'.

However, the plant's unscientific harvesting from the wild pose a considerable threat to its survival. Many commercially valuable rattan species have become extremely rare in their original localities today.

These include Calamus rotang, Calamus dransfieldii and Calamus travancoricus.

Scientists are of the opinion that Calamus neelagiricus, C.vattayila and C.pseudotenuis deserve the status of 'endangered' on the International Union for the Conservation of Nature (IUCN) Red List.

Rattans in Tamil Nadu

Tamil Nadu is home to a number of rattan species of which Calamus rotang is found in all districts of the state. Other species grow only in the Western Ghats spread over the districts of Coimbatore, Dindigul, Kanyakumari, Nilgiri and Tirunelveli.

Due to the availability of superior quality rattans that grow along the Kollidam river, rattan furniture-making industries have been functioning successfully in Thaikkal village of Tamil Nadu's Nagapattinam district since antiquity.

But in Tamil Nadu, no sincere and serious effort has been made so far to conserve rattans sustainably. It is quite uncommon for the state forest department to raise any cane plantation. In the research wing, some attempts have been made in the past to raise certain cane species on trial basis.

In the Mukkombu and in the Sholapuram research centres, some serious attempts were made by the research unit for raising Calamus rotang, C. thwaitesii and C. vattayila about a decade back.

Of the three species, only Calamus rotang performed well. But, unfortunately, no useful information about the cultivation of cane was evolved, resulting in abandonment of the cane trial plot.

About a decade back, the Palni Hills Conservation Council (PHCC) had raised rattan nursery near Dindigul. After visiting the nursery, I suggested to the divisional forest officer of Dindigul to attempt a plantation by using the seedlings as the division had suitable areas for such species. But, due to paucity of funds, it could not be carried out.

The forest department must develop important agro-forestry models for the proper growth and development of rattans. The department can then use the same technology for raisisng large scale plantations. Private farmers too must be supplied with suitable technology for rattan farming with suitable tree species and other silvicultural practices.

In the Western Ghats, large scale plantations of rubber, coffee, tea, cardamom, clove, and coconut have been raised. Each holding has defined boundaries such as streams.

In these private holdings, rattan species can be cultivated along the boundaries and stream margins. Rattans can be introduced to supplement the income of farmers and planters even in areas where they are not usually found

In countries like Philippines, the agro-forestry system has become successful with tree species such as coconut (Cocos nucifera), arecanut Palm (Areca catechu), jack (Artocarpus heterophyllus), custard Apple (Annona squamosa) and others. Agro-forestry has also been successful in other Asian countries like Malaysia, China and Indonesia.

Rattans have been a perennial source of income to the planters.

According to Innovius Research estimates, the bamboo and crane craft market in India was estimated at Rs.3,910 Crores in 2016.

As the extraction, conveyance, curing and other activitie are mostly done manually, the industry generates employment opportunities for rural people.

The cane furniture industry also does not require heavy investments. It can also play a major role in earning considerable foreign-exchange. The requirement of tree species for making furniture can be reduced to a great extent when vast areas of cane plantations are raised. When tree felling is reduced, the ecology of the area can also be restored to a considerable level.

It is high time that the Tamil Nadu Forest department rise up to the occasion to manage the economically, ecologically and commercially important rattan species sustainably for the betterment of the globe as a whole.

V.Sundararaju is President, the Society for Conservation of Nature, Trichy, Tamil nadu and Consultant with the Society for Social Forestry Research & Development, Tamil Nadu.

Rattans found in the Western Ghats of Tamil Nadu

Courtesy: Google

https://www.downtoearth.org.in/blog/tamil-nadu-must-sustainably-manage-its-rattans-71764

Monday 15 June 2020

37

INDIA MUST PROTECT ITS RARE, UNIQUE AND ENDANGERED PLANTS AND TREES

A systematic species recovery programme is the need of the hour

Syzygium travancoricum, an economically important tree species, is reported to exist with a population size of only 15-20 individuals.

Photo: Wikimedia Commons.

India is known for its rich biological diversity due to the presence of large numbers of plant and animal species. It is one of the top-ranking, mega-diversity countries of the world.

Our cultural diversity has played a major role in conserving the floral and faunal diversity. Having said that, this diversity is now in danger.

Take, for instance, trees. Despite their valuable services to humanity, trees are being ruthlessly destroyed because of developmental projects and increased dependence.

While several species are facing threats from anthropogenic pressure, many are threatened due to invasive alien species and climate change.

It is high time that the state and the central governments come forward for the recovery of the rare, endangered and threatened (RET) tree species.

Before taking scientific measures to ensure their conservation and cultivation, it is very much necessary to identify them, assess their natural distribution and study their population status.

Pristine forests have been fragmented because of intense developmental activities over the past few decades. More than 100 tree species of high economic importance have become threatened and critically endangered in the Western Ghats.

Their small population size is considered to be the major threat. It cannot sustain them due to inbreeding and loss of genetic variability. It has become an urgent need to carry out scientific conservation programmes for recovering these species.

Unless urgent measures are taken to restore the critically endangered species, it will become extinct. Species recovery is the process through which the decline of a threatened species is arrested or reversed and threats removed so that the survival of the species in the wild can be ensured.

Many countries have initiated plans to address the resurrection of the RET species. Species recovery programmes have been carried out successfully in the United States, Canada, United Kingdom and Australia.

In the United States, there is special legislation such as the Endangered Species Act, 1973, (ESA) for carrying out species recovery programmes. The Act, which was implemented in 1973, has provisions for listing the species as 'endangered', developing recovery plans for each species and designating critical habitats.

So far 47 species have been stabilized through different recovery processes and have been excluded from the recovery programmes. A gradual increase in the population size, habitat restoration and captive breeding or population stabilization have been achieved through recovery programmes.

Tree species in the Western Ghats

India's 1.34 billion people exert heavy pressure through encroachment, raising commercial plantations and other developmental activities. Due to anthropogenic activities like excessive harvesting and habitat destruction, many of the economically important tree species are under serious threat.

Syzygium travancoricum, an economically important tree species is reported to exist with a population size of only 15-20 individuals. Similarly, Dipterocarpus bourdillonii, another endangered species has only 14 individuals occurring in three patches in Kodagu district of Karnataka state.

Since tree species require decades for regeneration of the optimum population, if there is lack of regeneration or habitat, their present population cannot be considered healthy.

Increased inbreeding because of limited pollen and seed dispersal flow caused by fragmentation of populations can impact regeneration of the species. In case of Dysoxylum malabaricum, an endangered tree species in the Western Ghats, inbreeding between related individuals has caused reduced regeneration.

Hence, if urgent action is not taken to restore the population of these species, they may be irrecoverably lost. In India, recovery programmes for a few plant species have been taken up.

For instance, Paphiopedilum druryi, a slipper orchid, has been multiplied through tissue culture and has been reintroduced in the Agasthiamalai hill ranges of south Western Ghats.

Out of the 387 Indian plants listed under the International Union for Conservation of Nature's Red List, 77 have been enlisted as 'critically endangered', six are 'extinct' and two are 'extinct in the wild'. The IUCN is an international organization working in the field of nature conservation and sustainable use of natural resources.

The 77 critically endangered species can be prioritized for recovery programme and the balance can be taken up subsequently. A systematic species recovery programme is thus the need of the hour to restore the populations of these species.

While taking earnest measures for conservation and recovery of RET species, collection of data about the population sizes, identification of the specific threats and developing mitigation strategies are to be attempted systematically.

The Vallanadu Black Buck Sanctuary and the Grizzled Giant Squirrel Wildlife Sanctuary in Tamil Nadu; Aghanashini Lion-tailed Macaque Conservation Reserve in Karnataka and Gibbon Sanctuary in Assam have been established for conserving specific taxa.

But, till date no area has been specifically protected for any single endangered plant species, except for species groups like the Varsey Rhododendron Sanctuary in Sikkim and the Sessa Orchid Sanctuary in Arunachal Pradesh.

Combined and collective efforts are required on the part of the Forest department as well as the forest stakeholders.

Promulgation of specific acts and framing of rules and regulations are urgently needed to protect the threatened species by reaching an agreement between the department and the stakeholders.

Documentation of the RET species, their threatened status, surveying all known populations and mapping their locations, identifying the extrinsic and intrinsic factors that drive the species to threatened status and assessing the genetic variability of the species are to be carried out scientifically on war footing.

Based on the above strategies, long-term monitoring programmes are to be developed for assessing the population changes periodically.

https://www.downtoearth.org.in/blog/india-must-protect-its-rare-unique-and-endangered-plants-and-trees-72063

Wednesday 01 July 2020

38

TRICHY'S ELEPHANT RESCUE CENTRE IS A HAVEN FOR ABUSED PACHYDERMS

Since its commencement last year, the centre has cared for those elephants that have been ill-treated by private owners.

Since the commencement of operations last year, Trichy elephant rescue centre has become a refuge for many ill-treated elephants.

Photo: Tamil Nadu Forest Department

Tiruchirapalli's Elephant Rehabilitation and Rescue Centre (ERRC) is located at Marama Reddy Palayam, abbreviated as M.R. Palayam, a reserve forest that is 35 kilometres away from the city, on the Trichy-Chennai highway.

The centre was started three years ago. However it became operational when the Madras High Court directed the Tamil Nadu Forest Department to confiscate the private elephant, Malachi in June 2019, when the pachyderm was found to have been made to beg on the streets.

The ERRC, developed at Rs.2 crore, is spread over 20 hectares. A life-size statue of a young elephant erected on the roadside near the entrance greets passers-by.

The centre has been provided with the necessary infrastructure required for the rehabilitation of elephants in distress. The ERRC accommodates and looks after the sick and stressed elephants with the due permission from the Central Zoo Authority of India.

Before the commencement of the centre, the private owners who were found to misuse the animals were fined, or their license was suspended according to the provisions of the Wildlife Protection Act, 1972. Now the animals which are abused by private owners such as temple elephants are given refuge in this centre.

The funds for maintaining the centre and the payment of animal caretakers, are provided by the state government.

A similar Wildlife Rescue and Rehabilitation Centre functions under the control of the Bannerghatta Biological Park in Bengaluru. However, it provides refuge not only for the elephants but also for all other abused wild animals.

The elephants that are given shelter in the ERRC at M.R.Palayam, are provided with the best health care facilities in a natural environment. The centre is located in the forest about a kilometre away from the highway.

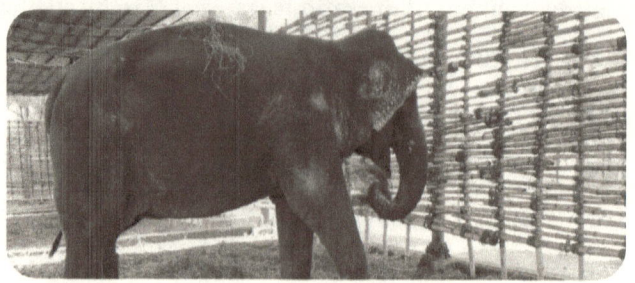

A bamboo kraal built for the animals at the centre

Photo: Tamil Nadu Forest Department

Unlike other centres, where visitors are allowed to watch and feed the elephants, the ERRC is meant purely for rehabilitation purposes.

The elephants at ERRC are looked after by mahouts and cavadis (Assistants to the mahouts), deputed from Anamalai Tiger Reserve (ATR). Every elephant is taken care of by one mahout and one cavadi. A veterinary doctor from the Tamil Nadu Veterinary and Animal Sciences University visits the centre weekly.

The elephants are taken to a human-made pond for bathing in the morning. They are then fed fresh fruits, vegetables, rice, barley, plantain, sugarcane and coconuts.

Afterwards, the animals go for a stroll through the 2600-hectare reserve forest. On completion of their walk, the elephants and their handlers take rest under a specially made shelter. Every day, the jumbos are given foot massage according to the instruction of the Doctor.

A large shower facility also has been created here at the centre. In addition to the grains, each elephant consumes 250 kilogram of fruit and vegetables daily.

The forest department has attempted to grow some food crops at the centre. A solid stable has also been developed here for the animals to rest and relax during the day time.

Two percolation ponds have been created to harvest and store rainwater. Residential buildings for the Mahouts and Cavadis have been readied. Departmental quarters also have been built for the field staffs that are in charge of the centre.

A veterinary hospital, kitchen, private walkway for the elephants and shed for the captive elephants have been constructed. The centre has the capacity of accommodating seven elephants at a time. As on date, the centre has given refuge for seven female elephants.

Closed Circuit Television cameras have been installed with special lighting arrangements for the safety of the jumbos. The maintenance cost of the pachyderms is expected to be around Rs.22, 000/day.

An elephant kraal has also been developed. Generally, the kraal is used to contain the restless behaviour of the newly-captured wild elephants. The bamboo structure will be strengthened, depending upon the requirement in future.

---

https://www.dpwntoearth.org.in/blog/wildlife-biodiversity/trichy's-elephant-rescue-centre-is-a-haven-for-abused-pachyderms-72377

Monday 20 July 2020

39
INDIA MUST CONSERVE ITS ORCHID WEALTH

Earnest measures are needed, involving awareness efforts among all stakeholders.

A rare orchid in India

The family 'Orchidaceae' constitutes one of the largest families of angiosperms. They are a highly evolved family with 600-800 genera and 25,000-35,000 species all over the globe.

Theophrastus (370-285 BC), known as the Father of Botany, gave the name 'Orchids' to these plants, based on the resemblance of paired underground tubers of the plants to male anatomy (the testes).

Because of this resemblance, people have the wrong notion that orchids possess aphrodisiac properties. These plants are perennial herbs with simple leaves.

Humanity has been familiar with orchids since the Vedic period. In the Rig Veda and the Atharva Veda (1500-800 BC), there is mention about Vanda tessallata (Rasna) and Flickingeria macrai (Sanjeevini) and their medicinal properties.

The Botanical Survey of India (BSI) has undertaken a study on 'Orchidaceae', only recently. There are 1256 species of orchids in India, the BSI estimates.

Of the 1,256 species belonging to 155 genera, 388 species are endemic to India. Of the 388 endemic species, about one-third (128) species have been found to be growing in the Western Ghats.

The Himalayas possess the maximum orchid species among the ten biogeographic zones of India. They are followed by North East, Western Ghats, Deccan Plateau and Andaman & Nicobar Islands respectively.

Orchids are divided into two groups, namely monopodial and sympodial. Monopodial orchids like Phalanopsis, Renanthera and Vanda grow continuously with central stem.

Sympodial orchids like Cattleya and Cympodium have a main stem that terminates growth at the end of each season. A new shoot that grows from the base becomes the bulbous stem known as pseudo-bulb which flowers finally. The pseudo-bulb stores food and water and functions like bulbs.

Orchids are classified into three categories namely epiphytic (plants that grow on another plant and rock boulders), terrestrial (plants that grow on land and climbers) and mycoheterotrophic (plants that derive nutrients from mycorrhizal fungi which are attached to the roots of a vascular plant).

Some 757 orchid species are epiphytic, 447 are terrestrial, and 43 are mycoheterotrophic in India. Since India has a vast number of orchids, hilly areas have some orchids flowering almost throughout the year. Grasslands also provide habitat for certain terrestrial orchids.

The minute seeds of the orchids have only minimal reserves of nutrients. As a result, the seeds depend upon mycorrhizal fungi for the carbon resources to germinate.

As orchids have mycorrhizal specificity, pollinator specialization, limited germination rates and sparse distribution in specific habitats, they are extremely susceptible to habitat disturbance.

Generally, the seeds require four to five years or more to attain flowering stage. The germination percentage is reported to be only about 0.3.

Under threat

Due to such reasons, the family consists of the maximum number of threatened species in nature.

The Orchidaceae family is under serious threat due to rapid destruction of natural habitat by deforestation, forest fire, overgrazing, felling of trees, rapid urbanization and indiscriminate collection for floral business.

Orchid species with medicinal properties are becoming very rare and endangered due to over-exploitation, coupled with lack of awareness.

The multifaceted adaptability and fast replicating characteristics of the invasive species like Eupatorium odoratum, Lantana camara, Parthenium hysterophorus, Ageratum conyzoides, etc., suppress the native flora including orchids in their original habitats.

Anthropogenic activities have thus become the major causes for the depletion of orchid wealth in Indian forests. Many orchids that were present in plenty in the past, have now become rare and endangered.

Because of its high value, endangered status and its significant role in the ecosystem, the family is often used as a flagship group in biological conservation. Orchids are considered as the indicators of the health of the forest ecosystem.

All the species of Orchidaceae have been listed in the Endangered Species of Wild Fauna and Flora in Appendix II of the Convention of International Trade (CITES).

What can be done?

Orchid diversity can be saved from extinction only if earnest measures are undertaken on a war footing for conserving their natural habitats.

The field staffs of the forest department across India have to be trained to identify orchids, as they are the custodians of their habitats. Experts on orchids should be engaged for imparting training to field staffs.

Till now, only Blue Vanda, Red Vanda and Ladies slipper orchids have been included under Schedule VI of the Wildlife Protection Act, 1972 restricting collection from forests. All the species of Orchidaceae family should be included in the Act for better protection.

Orchids must be protected through in situ and ex situ conservation for their long term survival in their natural habitats. Orchid seed banks and germplasm banks should be established.

By creating orchid conservation areas, tourists, college students and local people can be made aware of this plant wealth and its significance. With the support of the stakeholders, it may be easy to conserve this rare plant wealth.

Orchid sanctuaries can also be formed as is the case in Arunachal Pradesh and Sikkim. Arunachal Pradesh, also known as 'Orchid Paradise' is the first to establish the Sessa orchid Sanctuary exclusively for orchids. Assam is the second state that has set up Deorali Orchid Sanctuary.

---

https://www.downtoearth.org.in/blog/india-must-conserve-its-orchid-wealth-72884

Monday 17 August 2020

40
A UNIQUE REJUVENATION CAMP IN TAMIL NADU FOR CAPTIVE ELEPHANTS

Nutritious food and restorative care help relieve the stress that such elephants develop during captivity.

The Elephants Are Taken For A Walk After Bathing

Photo: Tamil Nadu Forest Department

According to Hindu cosmology, it is believed that mighty elephants support and guard the Earth. The deity Vinayaka or Ganesha, with the body of human and the head of an elephant is one of the most popular gods in the Hindu pantheon.

Elephants are featured in different religious practices. Especially in Kerala, elephants are used in temple festivals. Many temples in the state of Kerala possess their own elephants.

Elephants are also owned by some temples in neighbouring Tamil Nadu for use during religious ceremonies and other festivals. Many private individuals also own elephants and feature them in temple festivals and other functions.

Of the 29,000 elephants in India, about 2,500 are captive. But these animals are often not given proper diet and due care. As a result, they become aggressive and cause problems every now and then. Animal rights activists have protested the unfair and abusive treatment of the pachyderms.

With a view to providing these captive animals a new lease of life, former Tamil Nadu Chief Minister, the late J Jayalalitha organised an annual rejuvenation camp for a period of 48 days for the elephants owned by temples and other religious institutions in the state.

The camp was initially organised at Theppakadu in Mudumalai, Nilgiris District in 2003, 2004-05, 2005-06 and 2011-12.

As Theppakadu is quite far away, the venue was subsequently shifted to Thekkampatti near Mettupalayam on the banks of river Bhavani from 2012 onwards.

The camp is organised by the Hindu Religious and Charitable Endowments department (HR & CE). Generally, the camp commences in December and concludes at the end of January, after 48 days. The cost of Rs.1.5 crore for organising the event is borne by the State Government.

The camping arrangements are made by the Forest department. A six-acre piece of land within the reserve forest is provided with electric fencing all around, to avoid any possible intrusion by wild elephants.

Eight watch towers have been erected for monitoring the camp. Foresters, Forest Guards and Anti-Poaching personnel are deputed for regular monitoring and inspection of the camp. Necessary vehicles, torch lights, walkie-talkies and fireworks have been supplied to the officials by the HR & CE department.

Twenty-eight elephants participate in the camp. Of these, 21 are from various temples and five are from mutts. Two are from the temples of Puducherry.

Now the facility has been extended to the captive elephants of the Forest department as well. Twenty-seven each in the Mudumalai Tiger Reserve (MTR) and Anamalai Tiger Reserve (ATR), two each in Chadivayal elephant camp of Coimbatore district and Arignar Anna Zoological Park in Chennai and four elephants in the Tiruchirappalli elephant rehabilitation centre attend the camp at their respective locations.

The trained elephants were used in early years for timber operations by the Forest department. Now, as the areas have been declared as 'Protected Areas', the timber operations had been suspended completely.

Currently the elephants are used to drive away wild elephants that stray into farmlands and human habitations. They are also used for elephant safari and for patrolling activities inside reserve forest areas.

During the rejuvenation camp, the pachyderms are given complete rest, in addition to being provided with fruits, vegetables, sugarcane, plantain, horse gram, ragi, rice, salt, jaggery, vitamin tonics, mineral mixtures, etc.

A dedicated team of Veterinary Doctors and other assistants from the Animal Husbandry department is posted to monitor the health of the animals. The camp not only provides rest but also rejuvenation to the elephants both physically and mentally. Citizens are also allowed to watch the camping activities like feeding and training.

The elephants are diagnosed by the Veterinary Doctors and based on their recommendations, the concerned mahouts and kavadis take care of the animals round the clock.

The rejuvenation camp has helped the elephants to overcome the stress developed during captivity. Minor health problems faced by the elephants routinely are also taken care of.

In addition to health care and nutritious food, the elephants are given a good shower and a brisk walk twice daily for improving their health.

Animal lovers and the care takers of the elephants are strongly in favour of such camps as the pachyderms are made to undergo health screening and treatment for different ailments and other infections.

In Kerala, a month-long annual rejuvenation camp for captive elephants owned by private individuals is being conducted under the name of 'Aanayoottu'.

The Kanha Tiger Reserve of Madhya Pradesh too organises an annual rejuvenation camp for the captive elephants. Such rejuvenation camps should also be organised in other states that have captive elephants, to help the jumbos to relieve the stress and to maintain good health.

https://www.downtoearth.org.in/blog/a-unique-rejuvenation-camp-in-tamil-nadu-for-captive-elephants-73214

Thursday 03 September

41

WHY INDIA MUST CONSERVE ITS PALM TREES

The country has a variety of palm trees. But they are diminishing fast.

A pair of Talipot Palms (Corypha umbraculifera) in flowering near Tindivanam

Photo by Author

You all must have seen palm trees. They belong to the family 'Arecaceae' (the old name being 'Palmae').

Palms are an important component of the tropical forest ecosystem. The family also includes many species of economic importance.

There are about 2600 species of palms belonging to 200 genera all over the world. About 106 species of palms (with 22 genera) are distributed in India.

Palmyrah, Date, Sago, Caryota and many other palms are valued for their sweet water (neera), fruits, starch and other decorative plant parts. Palms attract the attention of horticulturists, foresters and nature lovers by their elegance and grandeur. The grace and splendour of palms probably made Linnaeus call them 'Princess of the vegetable kingdom'.

Palms of India

Palms are classified into three groups namely cultivated, ornamental and wild.

Among cultivated palms, the coconut tree (Cocos nucifera), known as Thennai in Tamil, is the most widely distributed species because of its ability to adopt varied climatic conditions. The coconut is called 'The Wonder Palm of Heaven' globally, because of its uniqueness.

The Palmyrah Palm (Borassus flabellifer) is known as Panei in Tamil and Pana in Malayalam. There are 8.50 crore palmyrah trees in India, of which four crore are in Tamil Nadu.

This palm is cultivated and self-sown throughout Tamil Nadu and Kerala. Until modern times, the folded sections of leaves, known as 'cadjans', were used in south India for letters, documents and records of all kinds, the writing being scratched on the surface with a stylus.

The sweet sap obtained from the cut ends of the peduncle is the main product. The sap is fermented into toddy or made into jaggery. Palmyrah trees have been an integral part of coastal India.

The tree is considered sacred and worshipped as Karpaga Tharu. It is the State Tree of Tamil Nadu. In spite of its multifarious uses, this palm is decreasing in number at a faster rate.

The betel nut palm (Areca catechu) is called Kamugu or Pakku in Tamil and Kavungu or Adakka in Malayalam. This is tall and erect tree largely cultivated in groves in Kerala, the Uttara and Dakshina Kannada districts and the foothills of the Nilgiris and other hill ranges of south India.

Its seeds (betel nuts) are used for chewing as they are aromatic and stimulant. The palm's spathes are used as plates, hats, etc. The betel nut palm is cultivated for its nuts that are chewed along with betel leaves and considered as an auspicious component of religious ceremonies.

The Kodai Panei or Koda Pana, also known as the Talipot Palm (Corypha unbraculifera) is a magnificent tree, with a height of about 24 metres. It is indigenous to the Andamans and Uttara Kannada. It is common to the Malabar coast.

The leaves are useful to make fans, mats, umbrellas, etc. The horny globose seeds are hardy, like ivory and are known as Bajurbet or Bazarubatu nuts that are made into necklaces, buttons and beads.

The pith yields a kind of sago used as food. This is 'monocarpic'. Monocarpic plants are those that flower, set seeds and then die. This tree flowers when about 40 years old, produces large number of seeds and dies in the ensuing hot season.

The Talipot Palm, when in blossom, bears the cone of golden-coloured flowers of about 15 feet high, which, when ripe, bursts with a loud noise and diffuses an unpleasant smell. Because of this pungent smell, the natives will not place their huts near the tree.

This is the most majestic and wonderful of the palm family. The stem sometimes attains the height of 100 feet.

Each of its enormous leaf may form a semicircle of 16 feet in diameter and cover an area of about 200 square feet on the ground. The 'Talipot' is as big and tall as a 'Ship Mast'. A single leaf will be able to cover about 20 men and keep them dry when it rains.

Palms like Bentinckia conddapanna, Arenga wightii, Pinanga dicksoni and Caryota urens have ornamental values.

Varei Kamugu or Kantha Panai (Tamil) or Chunda Pana (Malayalam) (Bentinckia coddapanna) is an interesting palm which occurs in the precipitous slopes of Kanyakumari district bordering Tirunelveli district, especially around Upper Kodayar.

Pockets of this species can be seen even from a distance and occupies the cliffs where the soil is very shallow. Along the depressions Alam Panei (Tamil) or Malam Thengu (Malayalam) (Arenga wightii) and Kana Kamugu (Pinanga dicksonii), are found in profusion.

Alam Panei yields highly intoxicating toddy that is much sought after by hill tribes. The nuts of Kana Kamugu are eaten by hill people as substitutes for betel nut.

Koontha Panei (Tamil) or Chunda Pana (Malayalam) is found growing in the forests of the Deccan and the evergreen forests of the Western Ghats. Its wood is used for agricultural implements, water conduits, buckets, rafters, etc.

The leaves are used as brushes, brooms and ropes. The starchy substance obtained from the inner part of the stem is made into flour and used as food. The spadix, when cut, yields toddy and sugar. A tree is said to yield 16 -80 cups of toddy a day. The leaves and stems are the favourite food of elephants.

Other Palms like Silver Date Palm also known as Peria Icham (Tamil) Phoenix sylvestris, The Date Palm (Phoenix dactylifera), Malai Icham or Siru Icha (Phoenix humilis) and Icham or Inchu (Phoenix farinifera) and Rattans such as Vanthai (Calamus brandisii), Kattu Perambu (Calamus rheedii), Calamus viminalis, Perambu or Churel (Calamus rotang), Chinna Perambu

(Calamus pseudo-tenuis) and Calamus viminalis are found growing wild in nature.

Diminishing fast

Till the middle of the 20th century, the Palmyrah Palm played a major role in the livelihood of a considerable portion of the rural poor population of Tamil Nadu. As the palm takes about 20-25 years to yield, it is dwindling drastically in numbers.

In spite of its immense uses, this palm is being destroyed in its habitats. Presently there are only about 12, 000 trees in the state.

There is an urgent need to revive not only this palm species but the entire family of Arecaceae.

The Kerala Forest Research Institute (KFRI) has established a 'Palmetum' that has live collections of indigenous and exotic palms. This centre is used for creating awareness about conserving palms among the public.

This Palmetum has a collection of 125 species of palms under 52 genera. There are 72 indigenous palms and 55 exotic ones. Such kind of Palmetum can be formed in Tamil Nadu as well, with the view of making people aware of conserving them.

Generally most palm species are found growing in forests areas. Many of them are becoming endangered due to habitat modification, encroachment and fragmentation.

Rare palm species should be brought under the Wildlife (Protection) Act and suitable measures should be initiated for protecting the same before they become extinct.

Despite the multifarious uses of this resource, significant attention has not been paid for conserving the same. It is high time that the Central and the concerned State Governments should take speedy measures for sustainable management of the same.

https://www.dowmtoearth.org.in/blog/wildlife-and-biodiversity/why-india-must-conserve-its-palm-trees-73343

Friday 11 September 2020

42
SUGGESTED TO INCLUDE THE WATERSHEDS IN THE H.A.D.P MAP

It all happened sometime during 1983 while I was working as the Forest Range Officer (FRO) of Udhagai North Range. There was a proposal for preparing a project report for Hill Area Development Programme (HADP) involving all the departments in the Nilgiris district. So, it was planned to visit all the respective areas along with the map prepared by the Agriculture Department. The Agriculture Department was entrusted with the job of writing the project with the support of all the other departments. The Heads of all the departments were requested to attend the inspection on a particular day by the District Collector. Accordingly, Officers from the concerned departments like Agriculture, Agricultural Engineering, Horticulture, Forest, Animal Husbandry, Health, Education, Highways, PWD, Panchayat, Tribal Development, Tourism, etc., assembled at the Collector's office. I was deputed to attend the programme on behalf of the District Forest Officer (DFO) of Nilgiris North Division. Around 10 am on that particular day, along with the Project Officer (DRDA), the team consisted of the officers from various departments started moving for inspection. The jeeps carrying the team members moved along Snowdown road and reached a place called Anikkorai, a village located on the northern side at the downhill of Snowdown Reserved Forest (RF), a hillock just above Raj Bhavan in Udhagamandalam.

The vehicles were stopped, and all the officers joined for discussion. Every department official was requested to discuss the possible aspects of developing the concerned area under HADP, so that whatever they propose, it may find a place in the proposal for getting sanction from the Government of India (GOI). One after one, it was discussed, and after consultation with other departments, without any overlapping, the salient points were noted down for incorporating the same in the proposal. When the turn for the Forest Department came, I pointed out the lapse noticed in the map, as it was prepared without including the hill slope of the Snowdown RF, which forms the main watershed for that particular area. Because, the Agriculture department had prepared the map for carrying out the activities by all other departments in about five villages located there at the foothill. Knowingly or unknowingly they had failed to include the hill slope which acts as the significant catchment.

Only if the watershed is treated properly, water security of the region can be ensured. As far as the Nilgiris is concerned, the shola forests act as the overhead water tanks of the district. Surprisingly the northern slope of the Snowdown RF facing Anikkorai village with degraded shola patches was the origin of a few jungle streams. Through proper soil and water conservation measures and revival of the existing degraded shola forests only, the forest ecosystem can be developed enabling to provide the ecological services like carbon sequestration, release of oxygen, rainwater harvesting, soil conservation, improvement of biodiversity, etc. After understanding the soundness of my argument, the PO (DRDA), the DD, Agriculture department and other officials agreed with me. So, as such whatever activities I proposed, they were noted down, and it was decided to finalise after discussing with the Collector.

Watershed or Catchment Area

At this juncture, a brief description about the watershed or catchment area may be of great help to the readers. A Watershed or Catchment Area is an area from which runoff due to rain flows into streams, rivers, lakes and finally reaches the ocean. It is also known as Drainage Basin. A watershed may be only a few hectares in case of small ponds or hundreds of square kilometres in case of rivers or major reservoirs. Since each watershed is an independent hydrological unit, any modification of the land use pattern in the watershed will have adverse impact on the water supply and the sediment yield. Watershed management involves optimal development of land, water and plant resources in order to meet the basic requirements of the people in a sustained manner.

Photo by Author

A watershed where runoff due to rain from the three sides (slopes) flows into the stream

The inspection went on till evening, and the team returned to the Collector's Office, Udhagamandalam around 4 pm. The Collector was available at that time. As everyone started taking leave, I too wanted to say goodbye. But the PO (DRDA) wanted me to accompany him while meeting the Collector. Though I was a little hesitant, he compelled me. PO, DD and AD, Agriculture department and I were present for the discussion. The Collector was eager to know the progress made that day. After discussing the various activities proposed by other departments, the PO mentioned about the lapse occurred in the treatment map prepared as pointed out by me. At that particular point, as the PO sought my help, I intervened and represented the key role played by the watershed in hilly regions like Nilgiris and the non-inclusion of the same in the map prepared. I also stressed the need for the treatment of the northern slope of the Snowdown RF, which acts as the major catchment for ensuring the water security. At that time the Collector was one Mr.Gupta. He agreed with my argument and instructed the officers of the Agriculture Department to revise the map, including different watersheds for the entire district and take my help while preparing the treatment map. Once the discussions were over, the Collector asked my name and details about me while taking tea in his chamber. I told my name and other information and expressed my helplessness while he was a bit unhappy with me for attending the meeting on behalf of the DFO, Nilgiris North division a few days ago. On that day as the Headquarters Forest Range Officer (FRO), I was deputed to attend a meeting with the Collector on behalf of the DFO. When I stood up to answer the query related to the Forest department, the Collector did not like anybody else to answer except the DFO, and as a result, I quit the meeting. I recollected the unhappy situation. After listening to my proper representation, the Collector said 'Sorry'. Subsequently, it became routine to meet the PO and the Collector for all the discussions about HADP.

Since the Forest Department plays a crucial role in Nilgiris district, due importance was given in subsequent events. Late Sri.H.Rajagopal Shetty.IFS, the then Conservator of Forests, known for his sincerity and vast knowledge, added many valuable inputs to HADP, and he was taken to Delhi for a few meetings in connection with HADP. Of course, we were able to carry out many useful works like the development of the degraded Shola forests, protection of Shola forests, consolidation of RFs, soil and moisture conservation works, etc., and the standard works carried out under HADP stand as testimony to my sincerity even after a long period of about 32 years. One will be surprised to see the intact check dam built during 1984 with water stored at Thalakundah and the cairns constructed opposite to the Hindustan Photo Films (HPF) around Muthinadu RF after more than three decades.

Thus, the entire ecosystem of the Nilgiris was developed to a great extent through the novel scheme HADP with the real involvement and dedication of devoted officers and of course which helped the Government to gain the goodwill of the stakeholders, as the project was extraordinarily beneficial to the local community.

Whatever work was proposed under HADP, it was aimed at to develop the ecology of the Nilgiris as the sustained and harmonious management of the same only could ensure the perennial supply of the ecological services that are considered to be very much essential for the wellbeing of all the living organisms on earth.

Checkdam built at Thalakundah in 1984 under HADP

(Photo taken in 2016 with the Author standing on the checkdam)

Photos by Author

Cairn built in 1984 under HADP, opposite to HPF in Nilgiris

(Photo taken in 2014)

43
CONTRIBUTION TOWARDS CONSERVING THE BIODIVERSITY OF THE NILGIRIS

Being part of the Western Ghats with highest rainfall and fertile soil, most of the RFs in the Nilgiris were infringed upon by the local villagers and the nearby people who own their lands. One such Government Forest is Ebbanadu RF which is situated about 18 km from Udhagai near Ebbanadu village. Most of the villagers own farms, but belong to the category of marginal and small farmers. As a result, naturally, they had developed a tendency of encroaching upon the nearby forest and cultivating crops. Otherwise, people are polite and friendly. One day when I perambulated the boundary of the Ebbanadu RF along with the Forester, FG and the FW, I noticed encroachments in many places and the same was verified with the RF map as it was my habit of carrying the RF maps whenever I inspected the RFs. The encroachments appeared to be very old, and without any detailed RF map, it would be challenging for the field staff to identify the infringement and take further action. There itself I instructed the Forester Mr.Ramalingam, (a sincere person from Nilgiris, who became FRO and retired) to take immediate action to evict the encroachments with due process.

As per my advice, one day the Forester prepared the eviction notices according to section 68-A of The Tamil Nadu Forest Act, 1882 and obtained my signatures and the same day itself the written announcements were served on to the encroachers at Ebbanadu in the presence of a person from the same village as a witness as the VAO was not available on that particular day. Since the number of encroachers was more and as we had to follow the procedure laid down under the above-said act, I went there and served the notices after organising a meeting with the villagers. Though there was an initial protest from the encroachers in receiving the notifications, after due persuasion, all the notices were served. Within five days of receipt of the notices, the encroachers had been directed to show cause and failing which, the encroachers would be summarily evicted from the said land.

We waited for about a week time and one fine morning planted Eucalyptus seedlings in all the vacant lands which were under encroachment when there was neither any agricultural or horticultural crop. The encroachers were very much shocked by our stern action. The next day, about 30 people from the village came to my official residence at Mount Stewart Hill, Udhagai. Luckily Mr.Ramalingam, Forester also was present at that time. The villagers pleaded with me to withdraw the planting in the evicted lands as they had only a minimal extent with which it would be tough for them to eke out their livelihood. After hearing their pleas, I told them very plainly that the evicted lands were parts of the Ebbanadu RF and at any point of time, they would be expelled as per law, and as such, it would be better for them to forget the land. Even a few persons started shedding tears, and I was helpless. The Forester also convinced them, and they left our residence with great disappointment.

Nearly after a fortnight, I received a summon from the District Court, directing me to appear before the Honourable Judge and depose in connection with the eviction of the encroachment in Ebbanadu RF. As per the court summon, on the due date I appeared before the Honourable Judge and gave my deposition with the served copies of summons under section 68-A of the Tamil Nadu Forest Act, 1882, Gazette Notification of the RF and the RF map showing the encroachments as exhibits and the present status of the RF. The Defence Lawyer after ascertaining all the facts and the action taken by us legally cross-examined me. About a week after, I received a letter from the District Court directing me to follow the procedure while evicting the encroachment.

After a month, one day, I received a call from the CF's office requesting me to come over there and meet the CF. At that time, Mr.Rajagopal Shetty, an eminent and knowledgeable Officer, was the CF of the Nilgiris circle. Immediately I rushed up and met the CF in his chamber. He inquired about the encroachment in Ebbanadu RF and the present stage after evicting the infringement. I explained everything to him elaborately and the representation made by the villagers in my residence also. He told me to give a detailed report about the long time encroachment and the effective action taken by us describing the different stages of eviction until the present status of planting. Being the Headquarters FRO, I used to meet the CF very often and explain all the difficulties whatever we faced during the forest management. So, already the CF had been informed about the various stages of the eviction process in Udhagai North Range. As per the instruction of the CF, a detailed report about the encroachments evicted in Ebbanadu RF with a map showing the long time encroachments was submitted to him on the very same day. After some time only I came to know that there was an instruction from the Chief Conservator of Forests (CCF), instructing the CF to transfer the Forester and me for the eviction undertaken.

The story behind the instruction of the CCF to CF was fascinating. The other day after meeting me in my residence, the people of Ebbanadu village had gone straight to the MLA. They had represented to him that they had lost their livelihood as the present FRO and the Forester had evicted their long time enjoyment of the forest land and sought a remedy. Subsequently, when the MLA happened to meet the Honourable Forest Minister in the presence of the CCF at Chennai, he had complained to the Minister that their party would be losing about one lakh vote because of the FRO of Udhagai North Range and the Forester as they had troubled the villagers of Ebbanad village. The MLA had insisted on transferring the Forester and me to some other place. The Honourable Minister had instructed the CCF to displace the

Forester and me very immediately. The CCF also, in turn, had asked the CF, Nilgiris Circle to transfer the Forester and me. But, the CF, instead of shifting us, had written to the CCF justifying the action taken by us to evict the long time encroachment and honestly protected us. It is challenging to see such kind of bold Officer nowadays in the Government service. When we do good work, only if we get the support of the higher officers, we can perform well. Similarly, only if we could gain the cooperation of the lower-level staff, we can do a better job. Thus we were duly rewarded for the excellent work done.

On a few occasions, forest fire would be in certain parts of the forests in the Nilgiris district, especially during the peak summer. Once I was informed by the Office Assistant (OA) of the DFO's office that there was a call from the Collector's bungalow about the forest fire noticed near Aramby RF on the way to the Collector's residence. Immediately I contacted the Fire Service Personnel who were stationed nearer to my house in the Mount Stewart Hill and requested their help. Next minute, they rushed up to the spot in a fire truck along with the fire fighters. The friendly relationship what I had developed with the Fire Service Officer (FSO) and the Fire Service Personnel was of immense use on such occasions. Those days we did not have the communication facilities as we have today. As the beat staffs were stationed on the south-western side of the Aramby RF, a hilly terrain, the fire occurred on the northern side of the forest might not be visible to them. After a few hours, the Fire Service Personnel met me in my residence and informed that they had extinguished the fire altogether. I expressed my sincere thanks to them.

Forest fire in the Nilgiris

Courtesy: Google

Next day the Aramby beat staffs would be informed about the previous day fire occurrence and requested to give a fire occurrence report. In turn, after knowing the excellent service rendered by the Fire Service Personnel, the beat staffs used to meet them and exchange their pleasantries.

Similarly, at times there would be a fire in Doddabetta peak. When I received the message, it would be informed to the Fire Service Personnel with a request to go over there with a fire truck and fire fighters to extinguish the forest fire. They used to oblige and immediately rush up to the spot. On the way, they would pick up the beat staffs that were staying in their quarters on the other side of the Doddabetta peak. After extinguishing the fire, if they returned very late in the night, they would meet me the next day morning and inform the matter. Next day the Doddabetta beat FG would submit the fire occurrence report.

Being a person with compassion and personal approach, I helped to extinguish the forest fire with the support of another department even without the knowledge of the concerned beat staffs. At the same time, there were

Officers, who had issued a charge sheet to the beat staff for not reporting fire occurrence within 24 hours. One should avoid that kind of sadistic attitude to perform the Government Service successfully.

Thus, to conserve the valuable biodiversity of the Nilgiris, I have contributed the maximum as FRO in my early years of service in the Tamil Nadu Forest Department, and I feel proud of it even now after more than three decades.

44
REVIVAL OF SHOLA FORESTS

Planting of exotic species was banned in the Nilgiris in 1987. Until then, Eucalyptus, Wattle, etc., had been planted routinely. Since there was an acute water shortage and as the people were against the planting of exotic species which was considered to be the cause for the water scarcity, the then Conservator of Forests (CF) Mr.V.R.Chitrapu, a generous and honest Officer took this decision. During that period, I was working as the Forest Range Officer (FRO) of the Hill Area Development Programme (HADP), with headquarters at Udhagamandalam. In 1985, we had planted Eucalyptus in about ten hectares in the Snowdown Reserve Forest (RF) as FRO, Udhagai North Range. Due to the sudden enforcement of the ban on the exotic species, we were forced to take up planting of the endemic species, i.e. the shola species, which are native to the Nilgiris.

Fortunately, we had raised about 5000 shola seedlings in Doddabetta RF. Naturally grown shola seedlings were collected from the existing shola forests at Doddabetta and transplanted into containers. When the seedlings obtained with the ball of earth intact and transplanted, the survival percentage was excellent. Almost we were able to get 100% success.

Doddabetta Shola Nursery

Naturally grown seedlings of the shola species like Berberis tinctoria (Nilgiri Berberry-Oosi Kala), Bischofia javanica (Bishop wood/Vinegar wood-Milachadayan or Cholavengai), Celtis tetrandra (Nettle Tree-Manja Pattani), Cinnamomum wightii (Wight's Cinnamon-Vettadu), Daphniphyllum neilgherrense (Soluvam), Dillenia pentagyna (Nai Thekku), Elaeocarpus oblongus (Nilgiri Mock Olive Tree-Vikki), Elaegnus kologa (The Wild Olive-Kulangi), Evodia lunu-ankenda (Kattu Shenbagam), Glochidion neilgherrense (Naidha), Hydnocarpus alpina (Malai Vattai), Ilex wightiana (The Holly-Vellodai), Litsea wightiana (Mashe, Keyanjee), Macaranga indica (Vattakanni), Mahonia leschenaultii (Holly-leaf Berberry-Mullu Kadambai), Michelia nilagirica (White Champak-Kattu Shenbagam), Microtropis ovalifolia, Neolitsea zeylanica (Tallow Tree-Shenbaga Palai), Photinia notoniana (Common Pulney Rovan-Kodibikki), Rhododendron nilagiricum (Billimaram or Poomaram), Rhodomyrtus tomentosa (Hill Gooseberry-Thavittu Koyya), Schefflera racemosa (Kanneemaram), Symplocos cochinchinensis (Sabal Tree-Bootha Kanni or Kambali vettu), Syzygium arnottianum (Vellanaval), Toona ciliata (Red Cedar-Malai Vembu, Sevvagil or Sandana Vembu), Turpinia nepalensis (Kanali or Nila), Viburnum erubescens (Narivele in Kannada), etc., had been collected and kept inside the Doddabetta RF under shade and maintained with great care.

We identified about 200 hectares in Snowdown RF to achieve the target fixed for the HADP Range for planting. As the terrain and the vegetation of the particular RF were familiar to me, I planned to take up planting on the north-western side of the RF. The Snowdown RF was having degraded shola forests interspersed with grassy patches. One day when we inspected the RF, we could notice about ten foot tall straggling spiny shrubs suppressing the younger recruits of shola species underneath. Throughout the RF, we observed many such sites. While we attempted to clear one such bush, we were surprised to encounter hundreds of younger seedlings grown underneath the thorny bush. So, we decided to assist the young ones by clearing the thorny shrubs, filling the gaps with the available shola seedlings and providing mulching for the newly planted seedlings in the entire area.

With the idea in mind, when I met the District Forest Officer (DFO) Mr.K.Chidambaram and discussed the matter, he suggested going for strip clearing. As he was busy with some other programme, he agreed to spare the Assistant Conservator of Forests Mr.V.K.Vadivelu on my request for field inspection. During the field inspection with the ACF, I showed him one such cleared patch with hundreds of young shola seedlings underneath. The ACF was very much surprised to see a large number of young ones. Then I expressed my idea of clearing the suppressing shrubs, reviving the recruits, filling up the gaps with the available shola seedlings in Doddabetta RF, providing mulching around the plants with available local material and providing nameplates for each species for identification purpose. The ACF was very much impressed with my ideas. After returning to the office, he explained everything to the DFO in a detailed manner. The DFO agreed wholeheartedly and permitted to go ahead with the planting programme.

Accordingly, we cleared the straggling shrubs without damaging the seedlings which had grown underneath. Shola seedlings were transported from Doddabatta and planted in the gaps noticed in the RF. The small branches and the leaves of the shrubs were used as mulching material and provided around the plants with raking up of soil to a depth of 15 cm and a radius of 20 to 50 cm depending upon the space available. The silvicultural operations right from planting were carried out with great care as the young recruits were fragile and tender. As the workers needed money, we made payment daily. The Foresters were very sincere, and the workers also were very cooperative as they all lived close to Snowdown RF. The people of Adasholai, the hamlet, located adjoining the RF, were very eager to see the forests in its initial stage.

While the planting work was in progress, one day the DFO visited the area surprisingly. On seeing hundreds of younger regeneration of shola species, he was very much surprised. He asked me what magic I did to have such a vast number of seedlings. When I explained everything to him, he was very much pleased and planned to bring the Conservator of Forests (CF) for inspection. Within a few days, he accompanied the CF for field inspection. The CF also felt happy by seeing a large number of natural regeneration and suggested to provide fencing all around the plantation as there were hamlets close to the RF. Clear-cut instructions were given by the CF to use the required number of Eucalyptus poles from the nearest coupes of South India Viscose (SIV) after treating them with the preservative supplied by the Forest Utilisation Officer (FUO) with chain link fence.

Sincere efforts were taken as it was the first of its kind to raise shola plantation in its natural surroundings. Chain link fence was provided after planting the treated Eucalyptus poles all around the planted area at an espacement of 2 metres. Shola seedlings, in general, are quick growing at the initial stage. So, naturally, the seedlings thrived with tremendous growth. Within a few months, an extraordinary increase was observed among the seedlings. The villagers were feeling happy by watching the fast growth of the seedlings and had unanimously decided not to allow any cattle inside the plantation. Some of them expressed their desire of seeing a panther in the patch of forest as they had seen the animal many years ago in the very same RF when the wood was in the prime stage. Therefore, it is evident that due to anthropogenic pressure only, the patch of shola forest in this RF had been degraded. That's why the CF had suggested providing fencing all around the plantation, it seemed.

Revived Degraded Shola Forest, Made as A Model Trial Plot

Within a period of six months, there was excellent growth among the shola seedlings and after a few more field inspections, the DFO wanted to make it a model trial plot for all the visitors and other dignitaries. Udhagamandalam was frequently visited by trainees from various forest training institutes all over India. Trainees from the Indian Forest Academy (IFA), Dehradun, Southern Forest Rangers College (SFRC) and State Forest Service College (SFSC), Coimbatore, Central Forest Rangers College (CFRC), Chandrapur and Northern Forest Rangers College (NFRC), Dehradun were taken to the plot and explained the activities carried out in detail for revival of the degraded shola forest and their role in harvesting the rain water and supplying the same to the people of Nilgiris in its purest form. Every House Leader was supplied with a cyclostyled copy of the detailed report about the shola forest prepared by me with due approval of the DFO. Those days, Xerox machines were not available. So, my report was typed and cyclostyled copies were prepared in the office. Photo by Author

Revived Shola forest of Snowdown RF. Photo taken during 2015 after 28 years

Though we were able to identify most of the shola species, for identification of certain doubtful species, the assistance of the Botany Professor and his associates of the Udhagai Government Arts and Science College, was sought.

They were also accommodating in identifying the species and nameplates were also provided to each species. Footpaths were developed across the plantation for taking up field inspection.

During one such field visit, as the bus could not navigate a narrow turn due to extended chassis of the limousine in which the Indian Forest Service (IFS) trainees from the Indian Forest Academy (IFA) travelled, they were made to walk about a kilometre to visit the plantation. By seeing the well-established shola plantation and the photos of the first degraded stage and the present stage of the vegetation, the trainees were impressed and inspired. As usual, they were supplied with biscuits and tea as a customary practice that prevails in the department from time immemorial. Impressed by the performance of the plantation, many trainees raised interesting queries to have more information about shola forest and its importance on their way back to the bus.

But, in general, the Toda Patta lands in the Nilgiris cause hindrance for the revival of sholas. A few shola forests are inside the Toda Patta lands. The Todas are not cultivating their lands. The lands are leased out for cultivation by non-Todas, a practice which is prohibited. They are extending the cultivation by clearing the shola forests resulting in shrinking of the valuable forest type.

The CF wanted to visit the Shola Forest revived at Snowdown RF once again. The date was finalised, and we were readily waiting for his arrival. The CF, the DFO and the ACF arrived and made a thorough inspection. The handout prepared about the plantation was shown to the CF, and the same was appreciated. The CF instructed to provide more number of nameplates for the well-grown seedlings and to keep the footpaths all over the plantation to take the visitors for seeing the entire area. The CF suggested to me to write an article on the Shola Forest as a co-author along with the DFO and send it to the Indian Forester, a monthly forestry journal published from the Forest Research Institute (FRI), Dehradun. He also suggested for keeping wooden staffs for the species with the idea of assessing their growth rate and recording them in the plantation journal. I produced the album containing photos with the initial stage of the vegetation and the present scene and different well-established shola species to the CF. I expressed to the CF that the work was made possible only by the support of the DFO through his technical knowledge. The inspection went on well, and the magnanimous CF requested the DFO to recommend my name for granting me an award for raising a shola plantation for the first time in the history of the Nilgiris. As per the request of the CF, the DFO also recommended my name and nothing could be heard of till I left the place.

Afterwards, there was a setback in the maintenance of the Snowdown shola plantation, it appeared. When our friend Mr.K.R.Varatharajan joined the NNFD as ACF during 1991, after inspecting the plantation, he had taken a great initiative to maintain it by repairing the fence, improving the path, etc. During his tenure in the division from 1991 to 1995, he had taken the Forest College Trainees to the plantation and explained the valuable services of the forest. He had not failed to mention my name among the trainees, and of course, I will be ever grateful to him for that act of kindness.

Anyway, without expecting any such reward or award, we carried out our activities sincerely.

Whatever we did, it was for fostering the ecological harmony and the results what we could see now after about a quarter of a century is heart-warming.

45
STRATEGIES FOR INCREASING THE BIODIVERSITY OF SIGUR PLATEAU IN THE NILGIRIS

Sigur Plateau is situated on the northern side of the Nilgiri Hills in the Nilgiris district. While the MWLS defines its west side, the Moyar River determines its north side with the Moyar Gorge to a depth of 260 metres. Moyar, Sigur, Avarahalla, Kedarahalla and Gundattihalla are the five major rivers that originate from the Nilgiris Plateau. Sigur Plateau is an important watershed area for the Cauvery River. Moderate and pleasant climate prevails in Sigur region. The temperature varies from 32° C in summer to 20° C in winter. The plateau which lies in the rain shadow region receives an average annual rainfall of 600-800 mm. Southern High-Level Thorn Forest is found in this region. Riparian Fringing Forests are located along the river and the stream banks. There are Montane Grasslands on some of the higher elevations of the plateau.

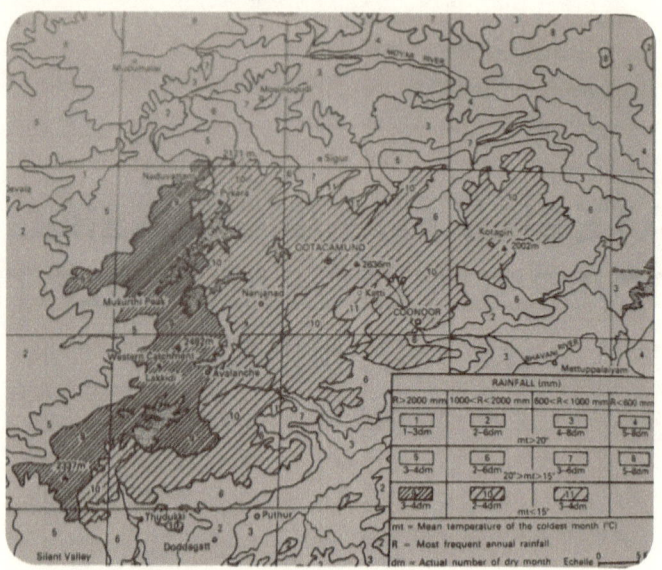

Bioclimatic Map of the Nilgiris

Flora of Sigur Plateau: Southern High-Level Thorn Forest found in the plateau at the foot of the Nilgiri Hills is a variant of Southern Thorn Forest type with numerous species of Dry Deciduous Forest type. Stunted specimens of Anogeissus latifolia and Soymida febrifuga occur scattered in varying numbers. The floristics of this variant is Chloroxylon swietenia (Satin wood or Porasu), Canthium dicoccum (Irumbaraththan or Naluvai), Zizyphus glabrata (Karikattan), Zizyphus xylophyrus (Kottai-elanthai), Santalum album (Sandal wood or Sandanam), Sapindus emarginatus (Soapnut or Puvankottai), Emblica officinalis (Amla or Nelli), Erythroxylon monogynum (Bastard sandal or Chembulichan), Capparis grandis (Mudkondai), Cassis fistula (Indian Laburnum or Sarakonnai), Cassia auriculata (Avaram), etc. Plenty of Sandalwood trees were grown naturally in this region.

In the Riparian Fringing Forest grown along the banks of the streams and the rivers, a few species of large trees, usually semi-evergreen are found typically forming a narrow fringe with smaller trees and shrubs between and often with a ground cover of coarse grass. Some of the tree species found here are Terminalia arjuna (Arjuna or Neer Maruthu), Mangifera indica (The Mango tree), Schleichera oleosa (Ceylon Oak or Puvam), Dalbergia latifolia

(The Rosewood or Eetti), Bambusa arundinaceae (The Thorny Bamboo or Perumungil), etc.

Faunal Population: The Sigur river valley supports a wide range of wild animals like Elephant, Gaur, Spotted Deer, Sambar, Mouse Deer, Barking Deer, Tiger, Panther, Wild Dog, Sloth Bear, Striped Hyena, Palm Civet, Small Indian Civet, Jungle Cat, Jackal, etc. As the Sigur Plateau acts as a wildlife corridor linking the forests of Sathyamangalam Forest Division being part of the Eastern Ghats and the forests of Nilgiri Biosphere Reserve being part of the Western Ghats, the area has the highest elephant population in India.

Elephant Corridors: Anaikatty Elephant Corridor, Masinagudi-Moyar Elephant Corridor, Singara-Masinagudi Elephant Corridor, Mavinahalla-Chemmanatham Elephant Corridor and Greater Moyar Valley Corridor are some of the vital Elephant Corridors of the Sigur Plateau.

Since the region contained a considerable population of Sandalwood trees and Elephants, illegal cutting and removal of the valuable aromatic trees and killing of the elephants for the valuable tusks were rampant in the 1980s and 1990s.

Rampant Smuggling of Sandalwood and Poaching of Elephants

As the sandalwood was renowned for aromatic and cosmetic value, it had very good rate in the neighbouring states like Kerala and Karnataka. This demand made the wood a precious commodity and its smuggling highly lucrative. The same thing happened with the elephant tusks also. "During 1983-86, at least 42 elephants were poached in Tamil Nadu with nearly a third of the cases from the Nilgiris North Division (V.Krishnamurthy in litt 1986) and Sukumar (1989) recorded 30-50 elephants being annually killed from the three southern states of Kerala, Tamil Nadu and Karnataka", as given in the book 'A God in distress'. Thus, the alarming level of poaching that resulted in a considerable decline of male-female ratio became the primary cause for dwindling of the elephant population.

During the year 1987 when I was working as the FRO in Udhagai HADP Range, these problems were at their highest peak in Sigur Forest Range which was under the control of the Nilgiris North Forest Division. Fortunately, we had sufficient funds for carrying out many forestry activities under HADP. One day during the monthly meeting, the DFO Mr.K.Chidambaram discussed this matter with the FROs of Udhagai North Range, Coonoor Range, Sigur Range and Udhagai HADP Range in the presence of the Divisional ACF Mr.V.K.Vadivelu. First of all, the reasons for such illegal activities in the RFs were discussed one after one. The main reason was found to be the lazy and passive attitude of the field staffs who were simply idling in Masinagudy without perambulating their forest beats. Though the FGs and the FWs were having their departmental quarters in different parts of Sigur Forest Range right inside the forest areas, due to lack of schooling and medical facilities, they were staying in Masinagudy village where both Sigur Forest Range office (Territorial Range) of Nilgiris North Division and Masinagudy Forest Range office (Wildlife Range attached to the MWLS) were located with their headquarters in the very same village. So, firstly the field staffs should be made to perambulate their beats.

Since there was no other field activity except protection, the field staffs were not showing any interest of inspecting the forest areas. If any forestry activity was created, then naturally they would be moving into the field. With this idea in mind, I suggested developing some kind of forestry activity like raising of miscellaneous plantation right from forming the nursery, construction of check dam, creation of percolation pond, etc., in vulnerable beats where such protection problems existed. Accepting my idea, the DFO in consultation with the ACF and the FRO, Sigur Range issued oral orders to raise plantations in three beats in an extent of ten hectares each, to construct a check dam in another beat and to create a percolation pond in a beat where the tribal people were facing water scarcity. Whatever discussed during the monthly meeting, written instructions followed on the same line for early execution of the field activities. As suggested, the FRO, Sigur Range was requested to issue a copy of the estimate to the concerned Forester, the FG and the FW so that there would be transparency of the activities and the fund allocated for each item of work. The field staffs were made responsible for the successful completion of the works on time. Automatically, after receiving the estimate copy for raising the nursery, plantation, construction of check dam and for creating a percolation pond, all the field staffs started visiting their beats regularly. The FRO, the ACF and the DFO also started inspecting the forests intensively and systematically. There was frequent perambulation of the forest areas by the staff which stopped the illegal movements of the poachers and other smugglers. The protection status of not only the sandalwood and the elephants, but also the entire biodiversity improved to a great extent within a few months.

We too were given about 20 hectares of plantation work in Sigur Range in addition to the other forestry activities in Udhagamandalam. Naturally, we started moving to Sigur frequently for raising the nursery and for selection of the planting site near Vazhaithottam in Sigur plateau. Fortunately, I had very good Foresters, namely Mr.Prem Kumar, who became ACF and retired and another person Mr.Sugumaran. Species which were suitable for raising in the region like Acacia leucophloea (Velvel), Acacia ferruginea (Parambai), Albizzia amara (Usil), Aegle marmelos (Vilvam), Azadirachta indica (Vembu), Chloroxylon swietenia (Purasu), Erythroxylon monogynum (Bastard Sandal-Cembulichan) and Zizyphus mauratiana (Ilanthai) were raised near Masinagudy. The area what we selected was having sparse growth of thorny species and a few trees scattered here and there. It was decided to fill up the gaps at an espacement of 3 m x 3 m. About 24,400 plants were raised at the rate of 1100 plants per hectare besides 2200 plants for casualty replacement.

After the onset of the northeast monsoon only during the end of October, we were able to take up the planting and complete it on time. Since we had some provision for providing solar power fence around the plantation as the plants were susceptible for damage by wild animals, it was planned to erect the same. As it is available nowadays, during those days this muchadvanced technology of building solar power fence was not there. So, we procured the energiser, steel wire and insulators from a private Engineer at Udhagamandalam. Stone pillars were purchased from Gundalpet of Karnataka state.

The fence was erected using the technical knowledge what we gained during our training on the solar power fence, and it started functioning. A 12-volt battery was used for the supply of electricity. The energy was boosted up through the energiser and supplied to the five strands of steel wire which provided repulsive force. So, when the cable was touched, it would throw away the live object. But, it would not catch the live object as it happens in case of regular electricity. In case of any wild animal coming in close contact with the electric fence, there would be a severe shock pushing away the beast without causing any injury. In the subsequent rains, the plants established well, and the solar power fence was also functioning normally. There was no damage to any plant by wild animals. It was often inspected by the ACF as the plantation was located close to the Sigur Ghat road which runs from Udhagamandalam to Mudumalai Wildlife Sanctuary (MWLS).

DFO testing the Solar Power Fence erected around Therkumalai East RF in Kanyakumari Forest Division during 2009.

Photo by Author

Similarly, in Sigur Range also, they had erected solar power fence in three places around the plantations. But, unfortunately, they could not succeed in their attempt and as it was expected.

One day I inspected the plantation along with the Forester and tested the functioning of the solar power fence. Long grasses collected there were used to check the pulse of the power supply along the solar power fence. On all the four sides of the plantation, it was tested. The functioning of the fence was found to be entirely satisfactory. The Forester was instructed to strengthen the room, built for keeping the battery and the energiser under safe custody. The Plot Watcher engaged for protection of the plantation was from the local village and was very sincere, and his father also happened to work in the Forest Department.

Then I proceeded to Masinagudy to meet the Divisional ACF. On my way, near Vazhaithottam, on the roadside, I could hear the familiar, loud and monotonous calls of some birds from the canopy of a banyan tree. As the tree was small, within a short while, I could identify the sparrow size birds as the Crimsonbreasted barbet or Coppersmith.

Crimsonbreasted Barbet or Coppersmith

The scientific name is 'Megalaima haemacephala'. The Tamil name is **'Semmarbu Kukkuruvan'**. This is a Sparrow size green colour bird with crimson breast and forehead. While the throat is yellow, the yellow underparts have green streaks. Both the sexes look alike. They are found in single or loose parties. Mostly they are found feeding on the fruits of the banyan and peepal tree. As their loud and metallic call is heard continuously resembling the hammering of a Copersmith on his metal, the bird is called 'Coppersmith'. They live in tree holes. They are omnivorous.

(Contd.)

Courtesy: Google

Crimsonbreasted Barbet or Coppersmith with a Ficus fruit in its bill

As I am passionate about forests and biodiversity, I enjoyed watching the birds for some time. Then I started moving towards Masinagudy. After reaching Masinagudy, when I was nearing the Sigur Forest Range Office, the ACF was discussing with someone that they would take the Deputy Conservator of Forests (DCF) from West Bengal (WB) to the plantation raised near Vazhaithottam by Udhagai HADP Range to show him the functioning of the solar power fence. The moment I entered the office, the ACF was thrilled to see me at the appropriate time and introduced the DCF, WB to me. As the ACF decided, the DCF was taken to the plantation near Vazhaithottam, and he was explained about the method of erection and functioning of the solar power fence. By seeing the thriving plantation in an area with plenty of wild animals, the DCF could understand the effectiveness of the solar power fence.

On completion of the planting and erection of the solar power fence, the DFO inspected the plantation one day. I and Mr.Premkumar, Forester were waiting for the arrival of the DFO. The DFO came around 11 am and inspected the plantation by walking to the four corners of the area. He had a glance of the Plantation Journal produced to him. The seedlings planted a month back had established well due to monsoon rain and no plant was damaged by any wild animal. He tested the solar power fence by using long grass and sensed the pulse. He was satisfied with the activities and asked the Forester to inform the species war numbers planted there. When I noticed the Forester looking with surprise and dismay for a moment, I opened my pocket notebook and read out the species war list with the numbers planted. By hearing the details of the seedlings planted species wise, the DFO could not say anything further. Then the DFO advised the Forester to learn the technique from me and left for inspection of other works in Sigur Range. But, the Forester was surprised to hear such accurate information from me at the right moment. With a smile, I told him that if I had not given the figures at that point of time, the DFO would carry a wrong impression on my administration. When I showed him the pocket notebook with some other information, he was taken aback and asked how I could give the numbers and the species correctly. I told him that it was due to my experience and dedication. I further said, "When a Superior Officer comes for inspection, we must be able to visualize what kind of information they may expect, and the figures and the names of the species were already inscribed on my mind".

In 1987, there was a programme of 'Creation of Ecological Farm' in selected divisions of the state. One such work was allotted to the Nilgiris North Division. The DFO told me one day in the presence of the ACF to carry out the work of Ecological Farm in Sigur Range. I was very much hesitant. I expressed my gratitude to the DFO for having a good opinion on my ability, and politely refused to do the work and listed out the reasons one by one. The area selected was within the jurisdiction of Sigur Range. When there was a Territorial Range Officer in charge of the Range, if we started doing the work in their jurisdiction, we might not get the required assistance. Conflict might arise, leading to unsuccessful performance. Whenever some work was done, if it was with the involvement of the local staff, we could gain the support of the local public and do the task successfully. So, I humbly requested the DFO not to involve me in that work. The ACF also endorsed my opinion. But, I promised to render all possible assistance from my side for carrying out the job successfully. Then the work of Creation of Ecological Form was done by the concerned Territorial FRO.

After a few months, the CF Mr.V.R.Chitrapu wanted to inspect the plantation and the solar power fence work done by Udhagai HADP Range at Sigur. That day the CF came along with Mr.R.P.S.Katwal, CF for inspection. Since the DFO was not available, the ACF accompanied them during the field inspection. By observing the performance of the plants, they were satisfied. Then they wanted to check the performance of the solar power fence. I told the ACF to touch the steel wire with long grass given to him. While he contacted, as he could not feel any electric current very immediately, he told me to check the line and the battery. I requested him to wait for a moment as it might take a few more seconds to complete the circuit. The very next moment, he felt the electric shock. Then it was tested by Mr.R.P.S.Katwal also. They were satisfied with the performance of the solar power fence and the seedlings planted.

When they proceeded to inspect another plantation raised by the FRO, Sigur Range in the adjoining area, the ACF requested me to accompany them as the concerned FRO was absent. During their inspection, by seeing some plants damaged by wild animals, Mr. R.P.S.Katwal asked me to replace the damaged seedlings with fresh ones. I nodded my head and said 'Yes, Sir'. While we proceeded further, he noticed some seedlings with exposed roots as they had not been appropriately planted deep in the pits. This time, Mr.R.P.S.Katwal became unhappy and pointed out the irregularities to me and instructed me to set right them immediately. This time also I gave the very same positive reply and told him very politely in a low voice that I was not the concerned FRO. In the meantime, our CF Mr.V.R.Chitrapu told the ACF to rectify the defects immediately and report the matter to him. As I had not served under Mr.R.P.S.Katwal, I was not familiar with him. But, at the same time, when we visited him in his residence during 2003 at Dehradun, while we underwent some training in the Indira Gandhi National Forest Academy (IGNFA), we were received very warmly and treated nicely both by him and his wife with apple, sweets, savouries and coffee. He was known for his benevolence in the department.

One day when the DFO, the ACF and I were having some official discussion, the DFO was feeling unhappy with me for not carrying out the Ecological Farm work as per his request, because the job was not done up to the standard as he expected.

Thus, different steps were taken in the Nilgiris North Forest Division to improve the forest cover and the biodiversity and wherever some devotion was shown, that particular operation had become successful.

46
ROLE OF FPS IN CONSERVING THE FORESTS

During 1999, I was working as the Assistant Conservator of Forests (ACF), Forest Protection Squad (FPS) with headquarters at Trichy. The erstwhile Trichy Circle consisted of Trichy, Thanjavur, Nagapattinam, Pudukottai, Dindigul and Kodaikanal forest divisions under its jurisdiction. Of course, now Dindigul has been made a separate circle with Dindigul, Kodaikanal and Karur territorial forest divisions under its control. So, as the jurisdiction was extensive with verdant and valuable forests, our job was also very hectic. While sandalwood was a significant problem in Salem circle, here in Thanjavur division, teakwood was the critical issue as far as protection was concerned. In Dindigul, there are magnificent forests as they are part of the Western Ghats. Quite a good number of valuable and naturally grown rosewood trees were found in the Reserved Forests (RF) and private forests also. Nagapattinam had been declared as a Wildlife Sanctuary, and Pudukottai division does not have good forests. As Kodaikanal is quite far away from Trichy, we used to visit the division now and then occasionally.

In this situation, one day, we visited Dindigul division based on a petition. Since it was a petition about some illicit fellings in Kaduguthadi RF, we planned to visit the RF for further verification. Our team consisted of two Forest Range Officers (FROs), two Foresters, two Forest Guards (FGs), one Forest Watcher (FW) and one Jeep Driver under the leadership of the

ACF. Our jeep reached Thadiyankudisai junction, and from there we decided to move towards Kaduguthadi RF. When we started driving towards the direction, our team noticed a boy coming from the downward slope towards the road. Mr. Thangaiya, Forester was asked to find out from the boy any clue about the illicit fellings as alleged in the nearby area as he had already served in the division. The Forester, along with one FG Mr. Kaliyamurthy got down from the jeep and moved downwards towards the boy. Since they were away at a considerable distance, what they discussed could not be heard. All the three then walked a little down and disappeared. Nearly after half an hour, they returned to the jeep and briefed that large scale fellings had taken place in the downward slope and requested all of us for the site inspection. When we got down and stepped on the ground, at the entrance of the downward slope, there was a small board fixed on a tree with the name of a person Kannayiram Pillai. We presumed that the particular plot of the forest might be owned by that person as mentioned on the name board.

When our team visited the said site, everyone was surprised and shocked by seeing such a vast extent of fellings. The tree felling site almost resembled a battlefield. Many trees had been cut and sawn by using massive size logs as supporting logs for sawing. Probably the operation would have continued for more than a month. Minimum of about ten truck loads of timber would have been removed. The unwanted small size logs, sawn planks and firewood had been left in the field itself as their removal would not be profitable. The waste planks and other wood left in the site alone might come to two to three truckloads. About fifteen naturally grown trees had been felled, and five of them were of enormous size trees. Our team members came to the conclusion that all the timber had been removed through the forest check post at Chitharavu under the control of Thandikudi Forest Range. The trees were of miscellaneous timber species including two Eugenia jambolana (Naval), two Acrocarpus fraxinifolius (Nellarai) and one Toona ciliata (Shevvagil or Santhana Vembu). The stump girths of these five trees were measured to be about 10-12 metres. Those days, for managing the private forests, there was special Ranger post as the FRO, Hill Preservation Act (HP) Range. Their main job was the implementation of the Tamil Nadu Hill Areas (Preservation of Trees) Act, 1955. As this area was under the jurisdiction of Thandikudi Range, it was under the authority of Thandikudi HP Range also in addition to the territorial jurisdiction of Thandikudi Territorial Range. After thorough field inspection, it was decided to inform the FRO, Thandikudi Range about the alarming situation found in his jurisdiction. So, we visited Thandikudi Range office situated at Vathalagundu (Batlagundu-old name) as its headquarters. The original name Vetrilai Kundru in due course of time got transformed into Vathalagundu. This is the main gateway to Kodaikanal. As we could not find the FRO there, it was informed to the field staff present in the office for further communication to the FRO.

That day itself based on another petition, we visited Sirumalai RF under the control of Sirumalai Range. The content of the complaint was that the field staff had permitted a private party to form a road through RF for transportation of quartz from a private land located inside the RF. During our field visit, the road was inspected by us along with the FG and the FW of the particular beat. It appeared to be a newly formed road, and we could notice fresh truck tyre marks. When we enquired the field staff, they told that the road was already in existence and the same had been used. They tried to explain that the road was not a new one, as mentioned in the petition. But our inspection and experience revealed that it was a newly formed road.

Then we reached the DFOs office at Dindigul, met the DFO and informed whatever irregularities noticed in the field. The Divisional ACF was also present there. I requested the DFO to spare the RF Map-Register, RF Gazette Notification and the Road Register for verification. I was informed that the RF Map-Register was missing. There itself, when the road register was verified, there was no mention about the newly formed road in Sirumalai RF, and it was confirmed that it was a newly formed one. On further verification of the Gazette notification of the concerned RF, there was mention only about a footpath existing across the RF. Therefore it was unanimously concluded that it was a newly formed road.

It was planned to survey the enclosure as mentioned in the Gazette Notification inside Kaduguthadi RF. The ACF was requested to come over to Kaduguthadi RF with the RF Gazette Notification and along with one FRO who was thorough with field survey on a convenient date as to be informed by our team. The DFO also gave clear instructions to the ACF to render necessary assistance to the FPS team and to tell Thandikudi territorial and the HP Range officials to be present during our next field visit. Tentatively it was decided to have a joint inspection of Kaduguthadi RF along with the Divisional ACF and the concerned field staff the day after as we had two more petitions to be enquired and verified. That night we stayed at Dindigul.

The next day we inspected Kannivadi HP Range jurisdiction in the forenoon. Though the FRO and the Forester of Kannivadi HP Range were informed well in advance about our field inspection, they did not turn up for investigation. In the hill area which was under the jurisdiction of the Range, we came across fresh fellings of rosewood trees in Kannivadi RF and in some private estates also. After felling and removal of the trees, the stumps were uprooted to a depth of about 30 cm in certain places and covered with soil. Similarly, in certain areas, the felled stumps had been burnt. We could enumerate about 50 such rosewood stumps. When we measured the stumps, the circumference of the stumps at the stump height varied between 1.5 and 3 metres. All those trees definitely would have excellent timber value. The felling list was prepared by Mr.S.V.G.Kuppuswamy, FRO of FPS and the same was verified by Mr.Dhandapani, FRO of FPS and signed by all the team members. The rosewood stumps burnt and covered with soil were also photographed for documentation purpose. Though in certain places even after covering with dirt, the felled stumps had sprouted and young shoots were noticed. As they were of a few months old, they were ignored. The young natural coppicing regeneration found in many places proved that this kind of clandestine operation was taking place regularly in that region. Such type of large scale fellings of the valuable timber tree definitely would not have taken place without the knowledge of the concerned Range staff. Our filed observation proved that the field staff had connived for such illegal activities.

Rosewood-Vulnerable as per the IUCN Red List

In order to conserve the endangered Rosewood tree species, a special act namely Tamil Nadu Rosewood Trees (Conservation) Act was introduced in the year 1994. As this tree species is endemic to the forests of Tamil Nadu, slow growing, dwindling due to indiscriminate cutting because of high timber value and there is every possibility of becoming extinct, the above act was enacted. In spite of the above said act in force, this tree species has become vulnerable as per the IUCN Red List. The above Act was amended in 2010 with due assent of the President on 12th February, 2010.

In the afternoon, we inspected North Eastern Slope RF in Oddanchatram Range along with the FRO and the section Forester in connection with the enquiry of a petition. Before that, we enquired the petitioner and recorded statement from him to the effect that the range staff had sold out a significant portion of the wind fallen rosewood tree to a private person and brought only a small part to Government account. The person who bought the rosewood also was enquired, and the statement recorded from him besides recovering the sawn size rosewood planks. The seized planks were handed over to the FRO, Oddanchatram Range for safe custody. The rosewood piece brought to the stock of the section Forester's timber register also was verified, and the entire tree was tallied. Required statements were also recorded from Section Forester and the FRO.

Virupatchi Forest Rest House

That night we stayed at Viruppachi forest rest house located next to Oddanchatram town on the way to Palani. Virupatchi forest rest house was our favourite halting place. Whenever we visited Dindigul Division, we had invariably stayed in the rest house as there was no demand for the rest house from any of the Government Official. Further, it was having a separate suit for my stay and the other suit was spacious enough to accommodate the remaining staff of the FPS team. Apart from this, it was easy for us to inspect Palani, Oddanchatram and Thandikudi Forest Ranges by staying here.

(Contd.)

Photo by Author

My Wife and me in front of the Forest Rest House at Virupatchi, Dindigul Division

As per the programme fixed with the ACF Dindigul division, next day we proceeded to Kaduguthadi RF along with our team. The ACF was present there at 9.30 am along with one FRO who was well-versed with the field survey with necessary survey instruments like field compass and chain. Other staffs of Thandikudi Territorial Range and HP Range were also present. The ACF had brought the Gazette Notification of RFs as discussed. Then we verified the field with the help of the Gazette notification. There was a private enclosure within the RF owned by an individual, namely Mr.Kannaiyiram Pillai. Certain salient features like old pathway constructed with a stone wall, naturally flowing stream, etc., were able to be identified. On the western side of the private enclosure, the stone boundary pillars were intact. The FRO started doing the survey, and the field book was prepared by the ACF. While doing the survey itself, we noticed that the five huge size naturally grown trees felled there were falling outside the boundary of the private enclosure. The ACF confirmed that the five trees felled were from the RF. It was reconfirmed by Mr.Dhandapani, FRO of our team. Fortunately, the ACF and Mr.Dhandapani were my juniors in the Southern Forest Rangers College (SFRC), Coimbatore. So, naturally, we three were able to take a unanimous decision about the irregularities. The trees had been felled within a period of about 2-3 months. The operation would have taken at least one month time. All the produces would have been transported through the forest check post located at Chitharavu only under the control of Thandikudi Range. One copy of the field book was prepared by Mr.Dhandapani, FRO of our team. After completion of the survey, a statement was recorded mentioning the irregularities noticed there in the RF, and the FROs, Thandikudi Territorial Range and HP Range and other concerned field staffs were requested to sign the statement. But the FROs and the field staffs refused to sign the statement. So, we got the signatures of the Divisional ACF, the FRO of the division who surveyed the area and the FPS team members including me. After having thorough field inspection with the Divisional ACF and the FRO, our team proceeded to Trichy via Dindigul. The ACF was requested to inform the DFO about the action taken by the FPS.

After reaching Trichy, we started preparing the report about the irregularities observed in Dindigul Forest Division and in the meantime; the matter was discussed in detail with the Conservator of Forests (CF), Trichy circle. After hearing the information patiently, the CF expressed his desire of inspecting the RF where such irregularities had taken place. The Dindigul DFO was informed about CF's inspection, and the field visit was fixed by the next day.

On that day, the CF was accompanied by the DFO, Dindigul division, FPS party, ACF and the concerned Thandikudi and HP Range staffs. He was taken to Kaduguthadi RF where fellings of naturally grown trees had taken place. By seeing such huge size stumps of the trees that had been felled in the RF, he was very much shocked. He was explained about the number of trees, approximate age of about 150 to 200 years of the trees and the stump girth ranging from 10 to 12 metres. As the CF was a Gold Medalist in Botany, he evinced keen interest to know the names of the species. Both the scientific and vernacular names of the trees were discussed in detail. Though he is a polite and soft officer of high calibre, by seeing such vandalism, he became angry and shouted at the field staffs. After inspecting the area thoroughly, the possibility of transporting the sawn timber was discussed by the CF with the DFO. Finally, it was confirmed that all sawn size planks would have been carried by trucks through Chitharavu check post only. Then the CF was taken by the DFO, Interface Forestry Division, Dindigul for field inspection in their division.

We returned to Trichy that day evening. The next day, the CF issued suspension orders to the FROs of Thandikudi Territorial Range, HP Range, Kannivadi HP Range and instructed the DFO, Dindigul to suspend all the other concerned staffs who were involved in the affair. In Oddanchatram Range, the Forester was suspended, and the FRO was charge-sheeted. So, a total of three FROs, four Foresters, three FGs and one FW were suspended in Dindigul Division. The situation was very alarming. The large scale suspension of field staffs sent shock waves to the entire state. A few officials who were connected with the Staff Associations contacted me to know the facts, and when I discussed in detail, they were all convinced.

Once during our night raid, when I met the DFO at Dharmathupatti check post, we two had a long conversation about the management of the division. Every forest division in the state has Registers of RF Map and RF Gazette Notification as they are permanent records. But, I was told that in Dindigul Division, the critical Register of the RF Map was missing. Some foul play was suspected in this regard. Whenever we visited Dindigul Division, I used to meet the DFO and have detailed discussion. Similarly, whatever irregularity was noticed, it was taken to the knowledge of the DFO very immediately for taking disciplinary action against the concerned staff either for commission or omission.

Thus the Forest Protection Squad (FPS) played a significant role in conserving the forest and the biodiversity.

47
SLEEPING GIANTS OF SIVAGANGAI

When the Tamil Nadu state has about 17.59% of the area under forest, Sivagangai district is unfortunate to have a meagre area of about 7.64% forest cover presently. According to the old records, the dense forest areas once occupied the district had been ravaged by the British while they attempted to capture the brave Maruthu brothers. Despite the sparse growth of the forest, the region is bestowed with certain rare and endangered tree species like Adansonia digitata, Manilkara hexandra, Spondias indica, etc. One age-old Manilkara hexandra tree is found well protected near Karaikudi. The tree Spondias indica is located within the premises of the Railway Office close to the Sivagangai Railway Station.

500 Year Old African Baobab of Vedhiyarendal

There were three African Baobab trees in the district some time back. But presently only the trees which are found growing in Vedhiyarendal near Manamadurai are well protected, that too because of the sanctity attached to the Village Deity Dharma Munishwarar. The other tree found on the bund of Ponnakulam Kanmai has vanished in recent years. The two sleeping giants at Vedhiyarendal are assessed to be about 500 years old. The two trees of the same species might have fallen due to some natural disaster centuries back. But as they are considered to be the toughest warriors who can survive the odds, the trees have survived now and a branch of a tree also has grown vertically to a height of 12 metres. The girths of the trees vary from 6 to 7.8 metres. The ancientness of the trees can be easily identified by watching the photos. The trees have spread over and covered an area of about 6000 square feet. Thanks to Lord Munishwarar for protecting the Organic Monuments.

Photo by Author

Adansonia digitata-Anai Pulia Maram or Pontham Puli- found in Vedhiyarendal DharmaMunishwarar Temple near Manamadurai of Sivagangai District-the tree found on the left side is the growth of a branch

Baobab-Ancient Tree of India

There is a mention about the Baobab tree in Chapter 15 of the Bhagvad Gita. Though it is mentioned as Banyan tree, the 'Upside-down' nature of its character strongly confirms that it is only the Baobab. When Lord Krishna talks about the Yoga of the Supreme Person, he said, "It is said that there is an imperishable Banyan tree that has its roots upward and its branches down and whose leaves are the Vedic hymns. One who knows this tree is the knower of the Vedas". The statement shows the ancientness of the Baobab tree in India.

Photo by Author

700 Year old African Baobab tree in the Chinmaya Vidyalaya School, Rajapalayam, Tamil Nadu

The majestic African Baobab is one of the longest living and the largest trees in the world. It is one of the most amazing trees. This belongs to the family 'Bombacaceae'. The native tree species of African continent dominates the dry, hot savannahs of sub-Saharan Africa. The tree is valued as a source of food, water, medicine and place of shelter. This is a rare tree in India, and one is amazed by the breathtaking grandeur of this mighty monarch with its extraordinary size and girth. The scientific name is Adansonia digitata. The common names are Dead-rat-tree (probably derived from the appearance of the fruit), Monkey-bread tree (the soft, dry fruit is edible), Upside-down tree (the sparse branches resemble roots) and Cream of tartar tree (cream of tartar). The common Indian names are Kalpataru, Kamalpati Vruksh, Rukhdo, Ghelo, Gorakh Imli (Hindi), Gorakh chinch (Marathi), Bukha (Gujarati), Brahmaamlika (Telugu), Gadhagachh (Bengali), Sarpadandi (Sanskrit) and Pondam Puli or Anai Puliamaram in Tamil.

The vernacular name 'Baobab' derived from Arabic means 'father of many seeds'. The scientific name Adansonia honours the French Explorer and Botanist, Michel Adanson who identified the specimen in 1749 in Senegal. 'Digitata' refers to the digits of the hand. The compound leaves with five leaflets (sometimes up to seven) are similar to a hand. There are eight species of the genus Adansonia. Baobab, a deciduous tree is a native of Madagascar (6 species), Africa (1 species) and Australia (1 species). Baobab trees generally grow as solitary individuals in savannah and scrubland. They are known for living over a thousand years. The tree is deciduous shedding their leaves during the dry season and remains leafless for nine months in a year. It grows to a height of 25 metres and 7 metres in diameter. The span of the roots exceeding the tree's height enables the tree to survive even in dry climatic condition. The tree is called 'Upside-down tree' as the trunk looks like a taproot and the branches are similar to finer capillary roots. The silver or grey coloured cork-like bark is fire resistant. The white, pendulous and showy flowers are enormous in size. The flowers bloom at dusk and wilt by dawn. While the flowers have a sweet scent initially, release a foul smell later when they turn brown. The flowers usually fall after 24 hours. Fruit bats pollinate the flowers primarily. The kidney-shaped seeds are hard and black.

Though this tree is native to Africa, it is also found in India, especially in dry regions and in Penang and Malaysia along a few streets. The tree is sensitive to water logging and frost. Its fruit contains calcium, vitamin C and high in antioxidants. The dry pulp is eaten fresh and dissolved in milk or water to prepare a beverage. The leaves are edible. The seed oil is used for cooking. In certain parts of Africa, the people soak and dissolve the dry pulp of the fruit and make juice. Elephants feed on the juicy wood below its bark. Animals like Elephants, Rhinoceros and Baboons help for the dispersal of the seeds. Baobab tree as it grows mostly in isolation attracts visitors due to the age, size, shape and extraordinary girth.

The Baobab tree is familiar in India for the past many centuries. It is believed that the African Traders had introduced the tree into our country. The trunk is known for storing a large quantity of water. The gigantic trees are known for storing water up to 1,20,000 litres inside the swollen trunks to overcome the harsh drought conditions. In some instances, the people use the hollow trunks of the living trees as houses. The trees may reach an age of over 5000 years.

A well-grown Baobab is capable of creating its own ecosystem supporting numerous creatures. Almost every part of the tree has medicinal value. The leaves rich in calcium, iron, proteins and lipid are used to cure inflammation. Its dried leaf powder is effective in treating anaemia, asthma, rachitis, dysentery and rheumatism. Its fruit pulp is known for curing dysentery, smallpox and measles. The fibre obtained from the bark is used for making rope, cloth, mats, baskets and paper. The seeds with vegetable oil can be grilled and eaten.

Baobab-a Magic and Sacred Tree

Baobab is considered to be the toughest warrior as it can survive the odds. It can be burnt or stripped of its bark or uprooted by storm. But it will just form a new bark and continue to grow. When it dies, it rots from inside and suddenly collapse into a heap of fibres. That is why people think that the tree does not die but simply disappears. These characteristic characters make it a '**Magic Tree**'. The African natives consider the Baobab as a '**Sacred Tree**', because of its exceptional vitality. There is a popular myth among the African people that if any prayer is done in front of the tree, the wishes would be fulfilled.

(Contd.)

Photo by Author

Though Nature has attempted to uproot the tree at Vedhiyarendal near Manamadurai, it has survived because of the mercy of Lord Dharmamunishwarar from the ruthless axe of wood cutters

The sight of the rare tree species may be something unusual for a Naturalist, Botanist, Ecologist, Scientist, Forester, and Horticulturist and even for a common man who has got the real attachment towards nature. So, everyone must protect such '**Living Monuments**' as they create a unique ecosystem providing shelter to numerous creatures, releasing oxygen, sequestering carbon dioxide, harvesting rainwater, conserving soil and supplying medicines for our ailments. The whole area at Vedhiyarendal with the spread of the trees including the idol of the deity Munishwarar can be declared as a protected area by providing suitable fencing in consultation with the local body and the temple authorities. Necessary hoardings can be erected to generate awareness among the people and other visitors about preserving such '**Heritage Trees**'. The Forest Department can play a lead role with the support of the District Collector and declare the area as protected one with the idea of conserving such rare and monumental tree species.

A **'Heritage Tree Conservation Committee' (HTCC)** has to be formed in every district to conserve the heritage trees. Already, it has been discussed in Chapter 1 in detail.

Heritage Tree Conservation Centre (HTCC) can be established in every district with the view of conserving the Heritage Tree identified in the regions. This will help in creating the necessary awareness to increase the tree cover throughout the state when there is not much scope for expanding the forest cover due to non-availability of the required land.

The tree cover expansion can be achieved only with the wholehearted involvement and support of the landholders, especially in the rural areas. Unless we have one-third of the land area under forest and tree cover, we can't attain economic stability. So, to have sustained economy and to get all our basic requirements fulfilled without any difficulty, the forest and tree cover has to be increased, and the establishment of such HTCC may be of great support in achieving **ecological harmony**.

Therefore, it is high time that the authorities concerned should rise up to the occasion to take earnest measures to protect such **'Living Fossils'** not only for the present generation but also for the posterity.

48
FUND GOT SANCTIONED BY THE NBM FOR CREATING SHELTERBELT

The Sri Sankara Educational and Charitable Trust, Tiruchirappalli had organised a seminar on Bamboos in Hotel Sevana at Trichy sometime in October 2005. I had gone over there to attend the workshop as I had been invited. I met the Head of the National Bamboo Mission (NBM), who happened to be the Inspector General of Forests (IGF). I had an excellent opportunity to interact with him. During our interaction, I explained about the various research activities carried out by the Industrial Wood Research Division of the Research wing. He mentioned that the main objective of the NBM was to increase the area under bamboo plantation in non-forest Government and private lands to contribute towards resilience to climate change. He asked me about the possibility of raising a shelterbelt using a few bamboo species. As we had raised shelterbelt at Dhanuskodi near Rameswaram, I had an elaborate discussion with him. I discussed the fourteen bamboo species grown at Mukkombu and invited him to have a look at the centre. Since he had to board the flight that day evening, he expressed his inability to visit the centre. I told him about the mangrove nursery raised on the shore of Bay of Bengal and suggested my idea of establishing a shelterbelt as per his desire by using a few bamboo species like Bambusa nutans, Bambusa balcooa, Bmbusa tulda and Bambusa vulgaris along the seashore of the Bay of Bengal with their support. He appeared to have been impressed with my discussion on shelterbelt, bamboo species, mangrove nursery and various research trials spread over the southern zone of the Tamil Nadu state. He requested me to send a proposal immediately for sanctioning the required amount for carrying out the work on time for successful completion.

As per the advice of the IGF, a proposal including a write-up, sketch and estimate was prepared for raising a shelterbelt plantation using Bambusa nutans, Bambusa balcooa, Bambusa tulda and Bambusa vulgaris mixed with other miscellaneous species along the seashore of the Bay of Bengal in Kodiyampalayam RF of Nagapattinam district and submitted within a week time. It was planned to use the tree species suitable for raising coastal shelterbelt plantation such as Casuarina equisetifolia, Casuarina junghuniana, Eucalyptus tereticornis, Anacardium occidentale, Acacia planiferons, Ficus hispida, Manilkara littoralis, Hibiscus tiliaceous, Baringtonia asiatica, Lannea coromandelica, Leucaena leucocephala, Cassia fistula, Morinda citrifolia, Thespesia populnea, Pandanus tactorius, etc., along with the other four species of Bamboo. Subsequently, a copy of the same was submitted to the CF, Research, Chennai.

Within a month, I received a letter from the NBM, New Delhi, along with the sanction order of the estimate sanctioning an amount of Rs.10 lakhs for establishing the shelterbelt plantation using bamboo species. Immediately I spoke to the CF, Research, Chennai over the phone and conveyed the matter. He was surprised to hear the information and told me that the proposal had to be routed through the PCCF only. But, on hearing that the estimate had already been sanctioned by the NBM, he felt happy and decided to discuss the matter with the CCF, Research and the PCCF.

By next day, the CF spoke to me that if the sanctioned amount was received by the Secretariat at Chennai, it would reach us very late. He wanted to evolve some kind of modus operandi to receive the sum directly so that the work could be carried out successfully.

He instructed me to select the site for establishing the shelterbelt and show it to him in the next camp. He advised me to raise the nursery also simultaneously. Suitable instruction was given to the FRO, Jayam Kondam Range as Kodiyampalayam RF was within his jurisdiction for carrying out the works on a war footing.

In the meantime, the CF one day informed me over the phone that necessary arrangements had been made so as to deposit the fund allotted by the NBM with the Forest Development Agency (FDA), Thanjavur Territorial Division. It seemed that a letter also had been written to the NBM by the PCCF requesting to deposit the fund with Thanjavur division. The concerned FRO could draw funds from the DFO, Thanjavur by submitting the required fund application. This kind of modus operandi had been planned to follow without any hitch in getting the funds.

As the proverb says, 'Man proposes; God disposes of', I was suddenly transferred to Dharmapuri as the Soil Conservation officer, Mettur Soil Conservation Scheme to the surprise of everyone.

But, any way I felt happy that during my short tenure in the Research Wing, I was able to get funds from the NBM through my efforts for establishing a shelterbelt along the coastal zone of the Bay of Bengal for protecting the inland from the storm, cyclone and tsunami caused by the changing scenario of global warming and climate change. Though shelterbelt is considered to be an artificial plantation, the high service rendered by it towards the protection of the natural ecosystem and the biodiversity is something incalculable.

Thus, I had the opportunity to use my knowledge and innovativeness for creating a coastal ecosystem that will foster the **ecological harmony** protecting the life of the inland people and the environment.

49

CONSERVING THE HILLS AND HILLOCKS OUTSIDE THE R.F

While I was working as the District Forest Officer (DFO), Kanyakumari District, this incident took place sometime during 2009. It was a Sunday. Since I used to be very busy during the weekdays, I could find time mostly on weekends only for my field inspection. That day I was heading towards Kodayar along the Nagercoil-Azhagiyapandipiuram road. While we were travelling somewhere near Erachakulam, a small village, casually I turned towards my left in the western direction. By seeing a few earthmovers (JCBs) on the top of a hillock, I was very much surprised. I told my driver to slow down the vehicle and look in the direction. He too confirmed the same. Immediately I instructed him to take the vehicle to the hillock. Within ten minutes, we reached the hillock. The extent of the hillock may be about 40 hectares. Of course, the hillock was about 3 kilometres away from the nearest RF. Two tipper trucks loaded with earth and one more with boulders were readily waiting for departure. By seeing that, I became a bit angry as a natural ecosystem was in danger. Though I tried to hide my temperament on my face, my talk with the concerned persons naturally revealed it. On my interrogation, one fellow showed me an order signed by the District Collector, allowing removal of earth and boulders from the specific survey number. But nothing was mentioned about the landscape as a hillock in the order. Our discreet enquiry revealed that three Doctors had bought the hillock with the plan of building a college after annihilating it by way of removing the earth and the boulders. There was terrific demand for soil and stones in Kanyakumari district, and they were mostly transported to the nearby Kerala state.

Immediately, I tried to contact the Collector over my mobile. His number was busy. Next, I contacted the Assistant Director of Mines. His mobile was not reachable. Then I spoke to the Tahsildar. When I enquired him about the permission given for the removal of earth and boulders by destroying a hillock near Erachakulam, he said that he had joined only a week back and yet to see the spot. In the meantime, Collector came online. When I narrated him about the order issued by him for the destruction of a hillock near Erachakulam village for construction of a college and the violations noticed in the implementation of the Hill Area Conservation Authority (HACA), he requested me to inform the proceeding number and date. I passed on the information as required. He said, being Sunday that day, he would verify it the next day and take suitable action.

The next day I had gone over to the Collector's office to attend a meeting. While I was waiting in the conference hall, one Office Assistant came and informed me that the Collector wanted me to come over to his chamber. I met the Collector in his room. After exchanging the usual pleasantries, Collector told me that he was not appropriately guided about HACA in that matter and hence had cancelled the order instantly.

There is one Dr.Lalmohan, a noted Environmentalist and Convener of the 'Indian National Trust for Art and Cultural Heritage' (INTACH) in Nagercoil. He is a good friend of mine. I discussed the matter with him, and within a week, he organised a seminar on the topic 'Conservation of Hills and Hillocks'. The workshop was to be inaugurated by the Collector and the keynote address to be delivered by me. Since the Collector had some urgent programme at Trivandrum on that day, I had the opportunity of inaugurating the seminar at the Notary Public Hall, Nagercoil. I have given the gist of my inaugural address in the box.

Hills and Hillocks-Important Terrestrial Ecosystems

"Hills and Hillocks become one of the important terrestrial ecosystems. We have to clearly recognize the difference between the hills and the hillocks. While Yercaud hills, Kolli hills, Sirumalai hills, Kodai hills, etc., are called hills, Thirupparankundram, Senkundram, Nerkundram, Pazhani malai (where Pazhaniyandavar temple is located), Viralimalai, etc., are hillocks. Though generally mountains, hills and hillocks are called **Malai** in Tamil, they have got their own difference. **In Tamil Nadu state, there may be about one thousand hillocks.** Hills and hillocks play a crucial role by providing valuable ecosystem services. With the available soil, air and water they support a variety of plants and animals including microorganisms. Thus, they not only provide shelter for various species of animals, birds, reptiles, insects and microorganisms through the grasses, herbs, shrubs, climbers, creepers and trees that have grown on them, but also absorb carbon dioxide and release oxygen, conserve the soil and water in addition to moderating the climate of the particular region. When we have hills and hillocks, the plant community on them harvest the rain water and store underneath. The rain water is gradually released as springs, finally taking the shapes of streamlets, streams, rivulets and rivers. These water bodies make the area fertile wherever they pass through. Eventually when they join the sea, they provide life to the marine organisms besides carrying valuable minerals with them. Only an Environmentalist, Ecologist, Naturalist or Scientist can understand all these intricacies. But, fortunately many hillocks are the abodes of Gods of various religions like Hinduism, Christianity, Jainism and Islam. Because of the spiritual protection, they are safe to a certain extent. We must be ever thankful to our ancestors in this regard".

(Contd.)

Inaugural address by the Author in the seminar organized by INTACH at Nagercoil

Hill Area Conservation Authority (HACA)

HACA is in vogue in many of the districts which are close to the Western and the Eastern Ghats in Tamil Nadu. One should take permission from the Director of Country and Town Planning for carrying out any developmental activity in such areas. Due permission is very much required even to construct a building of more than 300 square metres. The Committee under HACA headed by the Director as the Chairman has the Principal Chief Conservator of Forests (PCCF) and the Chairman, Pollution Control Board (PCB) as the members. Whatever application is received requesting permission from the HACA authorities will be forwarded to the District forest Officer (DFO) concerned for inspection and report.

During 2015, when I had gone over to Kanyakumari as the Consultant, Society for Social Forestry Research and Development Tamil Nadu (SSFRDT) for monitoring the activities carried out in Kanyakumari Forest Division under the Tamil Nadu Biodiversity Conservation & Greening Project (TBGP), I was happy to notice the hillock in the very same condition.

Even now, many hillocks have been vandalized by the greedy quarry contractors of the districts like Madurai, Kanyakumari, etc. Certain hillocks have been notified by the Revenue authorities as water bodies in a region like Madurai. For example, the Pancha Pandavar hillock, an archaeological site has been listed in the revenue records of Madurai district as a water tank in Keezhavalavu panchayat. Likewise, as the hillock with Jain stone beds dated back to 2nd century BC has not been entered in the Prohibitory Order Book (POB), this hillock has been quarried, even after repeated representations by Keezhavalavu Panchayat President to the Collectors.

Hence everyone must take sincere efforts to protect the hills and hillocks as they play a significant role in improving the environment around them and maintaining the **ecological equilibrium.**

Perumalmalai (A Hillock) in Thuraiyur   Sirumalai (A Hill) in Dindigul

Courtesy: Google

50
SAVED SAKUNTHALAI MALAI

Sometime during 2009, there was a proposal to construct a dam by the Public Works Department (PWD) across Ulakkai Aruviyaaru which flows from the famous Ulakkai Aruvi waterfalls located in Asambu Reserve Forest (RF) of Kanyakumari Forest Division. This river forms a tributary to Pazhayaru, a famous historical river that has mention about it in the ancient Sangam Literature of Tamil as 'Paguruzhiyaaru'. According to the Historians, over a period of time, the name 'Paguruzhiyaaru' has been modified into 'Pazhayaaru'. The dam was proposed here to augment the water supply to Nagercoil town. Since the river originates and flows through Asambu RF, the PWD requested the Forest Department (FD) to allot 4 hectares of forest land from the RF for that project. As per the Forest Conservation Act, 1980, if a portion of any RF is to be denotified for non-forestry purpose, twice the extent of the land is to be given by the concerned department as compensatory land. The only condition is that the area which is proposed as compensatory land should be contiguous to any RF. In addition to this, the fund required for afforestation and for carrying out soil and moisture conservation measures is also to be shared by the concerned department with the Forest Department. After fulfilling all these conditions, due permission has to be obtained from the Government of India (GOI).

Ulakkai Aruvi Falls

This falls is about 17 kms from Nagercoil town. As it resembles the form of a pestle (a heavy tool with a round end used for crushing substances in a special bowl called a mortar with the Tamil name Ulakkai), it is called Ulakkai Aruvi Falls. The falls located in the Asambu Reserve Forest is accessible only on foot from the base of the hills. The trek may take about one hour through rocky and forested area. As the falls is within the RF, due permission should be obtained from the Forest Department. People are permitted to reach the falls, take bath and relax without spoiling the atmosphere.

Courtesy: Google

Ulakkai Aruvi Falls in the Asambu hills of Kanyakumari district

The PWD authorities had requested the Collector and got approval for allotting about 8 hectares of revenue land located close to Velimalai RF. So, accordingly, that land was shown to us. Since the area was having many encroachments, it was rejected. Kanyakumari district is known for infringements. If we accept such area as compensatory land, then the headache of evicting the invasion will fall on our head, and it will become a perennial problem. Senior Forest Officers cannot forget the ordeals faced by the Forest Personnel when they tried to evict the encroachments in the Jenmam lands of Gudalur taluk of Nilgiris district after accepting the areas from the Revenue Department with the intrusions existed. It is the bounden and foremost duty of the Senior Forest Officers to inform the younger officers about the intricacies involved in such situations to improve their management skills.

As the project had to be implemented early as there was acute water scarcity in Nagercoil town, the PWD authorities were in constant touch with the Collector for getting a suitable land to hand over the same to the Forest Department. In this situation, one day, I and the Collector were discussing specific issues in his chamber. At that time there was a phone call to the Collector. From the way he reacted, I was able to understand that somebody at a higher level had requested him for some obligation. Once the call was over, he asked me to suggest some way to save Sakunthalai Malai. I thought it over for a while and very immediately suggested to allot that hillock as compensatory land instead of 4 hectares of forest land from Asambu RF which was required by PWD for constructing a dam. I added further that as Sakunthalai Malai was close to Therku Malai West RF (a hill range which runs parallel to Aralvai Mozhi to Kanyakumari road on its eastern side), there would not be any problem in accepting the hillock. By hearing my suggestion, he became pleased. I was told that a so-called Educationist who hailed from the same district wanted Sakunthalai Malai on lease basis for quarrying as he needed stones for forming a private port there. Since that person was having many educational institutions in Chennai and had close contact with the politicians, the Collector was having the apprehension of some political pressure in the matter. As already the people who live close to that hillock and other Environmentalists had raised objection for quarrying and sent petitions to the District Forest Officer (DFO) and the Collector, it was decided not to allow any quarry there. So, immediately he called the Revenue Divisional Officer (RDO) and the Tahsildar concerned and discussed the matter in detail.

As most of Sakunthalai Malai has rocky outcrops, it was decided to allot 20 hectares to Forest Department retaining about 10 hectares with the Revenue Department as they may require the same for future development.

The Divisional Engineer (DE), PWD was informed immediately. As seven copies of the proposal had to be prepared with necessary maps of the RF and the compensatory land along with other details for submission to the GOI after signing them by the Collector, the DFO and the PWD Engineer, the Revenue authorities and the PWD Engineers were requested by the Collector to assist me in preparing the same for early submission. Sakunthalai Malai also was inspected jointly by the Revenue authorities, PWD Engineers and the DFO. After the field inspection, all the three department officials sat together tightly and prepared all the maps and filled up the forms meant for the same. That day evening, seven copies of the documents after duly signed by the DFO and the PWD Engineer were sent to the Collector for approving and submitting to the GOI. As the list of the standing trees had been prepared, and the value also had already been assessed, it was easy for us to fill up the forms.

An Able and Devoted Person of the Kanyakumari Forest Division

At that time there was an able and devoted person in the District Forest Office who had thorough knowledge about the encroachments and the various court cases that were pending with different courts. He was one Mr. Harihara Iyer, a retired Inspector of Survey. But, everyone used to call him 'Samy'. He was working in the department for more than two decades. He was engaged temporarily to assist the DFO, the Draughting section and the Lease Assistant in preparing counter affidavits for court cases, in evicting the encroachments, in preparation of proposal for recommending quarry and in processing papers under Hill Area Conservation Authority (HACA), etc. He was of great help in preparing the forms and maps which had to be submitted to the GOI. Even at the age of 78, he used to accompany us to the field for

> verification of the boundary and permanent marks of cairns of various RFs. He was more thorough with the boundary and cairn marks than the beat and the section staff. Similarly, he was thorough with the case files also. Once, while I was travelling to Chennai by train to attend a meeting in connection with pending court cases, when I discussed with him, he was able to share the details about different cases elaborately without any file. Recently he passed away and his absence is really a great loss to the department. Many times, the Collector used to take Samy's help while signing the counter affidavits related to forest cases in which he was the first respondent.

The Collector signed all the seven copies of the forms, and the maps and the same was submitted that day itself to the Government of India. Incidentally the next day I happened to be with the Collector in his chamber. When there was a phone call, he replied that the said hillock had been proposed as compensatory land to the Forest Department instead of the forest land required by the PWD for construction of a dam. He added further that as the file had been sent to the GOI, nothing could be done at that stage. By hearing his convincing words, the Officer could not say anything further and said OK. Subsequently, I was told that the person who spoke to the Collector regarding the quarry lease was none other than the topmost Officer of the Government of Tamil Nadu. It was evident that as the Collector expected this kind of heavy pressure from the higher up, he was in a hurry in submitting the proposal to the GOI.

Thus, Sakunthalai Malai, a hillock located contiguous to the Therku Malai West RF, was saved by our diligence and timely action. By way of protecting the hill, we were able to help the district as a whole to enjoy the valuable ecosystem services forever. Once when there was some discussion about the incident, Mr.Harihara Iyer recollected the sincere efforts of like-minded officers in overcoming a crisis in saving one of the precious natural resources.

The task was able to be achieved only because of the sincere and earnest teamwork of officers having similar opinions and interests. When such officers with optimistic ideas happened to work together, this kind of challenge could be overcome easily.

That too, when it is towards conservation of any natural resource, it is a great pleasure to carry out such activities enthusiastically and earnestly as humankind is also part and parcel of Nature.

Photo by Author

View of Therkumalai West Reserve Forest from different angles

WORKS CONSULTED

1. Balachandran, *Indian Bird Banding Manual*, Bombay Natural History Society, 2002.
2. Balachandran et al, Annotated Bibliography of Point Calimere, Tamil Nadu Forest Department, 2010.
3. Baruah.A.D, Muthupet Mangroves & Canal Bank Planting Technique, Tamil Nadu Forest Department, 2012.
4. Basappanavar & Kaveriappa, *Romancing Elephant*, Vanasuma Prakashana, 2007.
5. George B. Schaller, *The Deer & the Tiger*, Natraj Publishers, 1998.
6. J.Narayanasamy, Thirukkural, Tamil and English Version.
7. John Joseph.S & Sundararaju.V, *Management Plan for Mudumalai Wildlife Sanctuary*, Tamil Nadu Forest Department, 1978.
8. Kailash Shankala, *Tiger! The Story of the Indian Tiger*, Rupa & Co, 1978.
9. Kamba Ramayanam, a famous Tamil epic written by the Great Tamil Poet Kambar during the 12th century.
10. K. M. Mathew, *The Flora of the Palni Hills*, The Rapinat Herbarium, St. Joseph's College, Tiruchirapalli-620002.
11. Kerala Forest Department, *Silent Valley - Whispers of Reason*, 1999.
12. Kopmeyer, *Thoughts to Build on*, UBS Publishers' Distributors Pvt.Ltd, 2003.
13. Krishnan. M, *An Ecological Survey of the Larger Mammals of Peninsular India*, Bombay Natural History Society, 1972.

14. Gulf of Mannar Biosphere Reserve Trust (GOMBRT) Publication, *Mannar matters*, 2007.

15. Manimekalai, one of the Five Great Epics of Tamil Literature written by the Poet Saaththanaar.

16. Pattina Palai, Tamil Poetic work belonging to the Sangam Period.

17. Prater, *The Book of Indian Animals*, Bombay Natural History Society, 1971.

18. Purananooru, one of the eight books in the secular anthology of Tamil Sangam Literature.

19. R.Annamalai, Biodiversity of Kalakkadu Mundanthurai Tiger Reserve, Tamil Nadu Forest Department, Government of Tamil Nadu, Chennai, 2004.

20. Ravikumar et al, *100 Red-listed Medicinal Plants*, Foundation for Revitalisation of Local Health Traditions (FRLHT), 2000.

21. Rangaswami and Sridhar, *Birds of Rishi Valley and Renewal of Their Habitats*, Rishi Valley Education Centre, 1993.

22. Romulus Whitaker, *Common Indian Snakes*, The Macmillan Company of India Limited, 1978.

23. Salim Ali, *The Book of Indian Birds*, Bombay Natural History Society, 1996.

24. Satish Pande et al, *Birds of Western Ghats, Kokan and Malabar*, Bombay Natural History Society, Oxford University Press, 2003.

25. Sekar. T, *Forest History of The Nilgiris*, Tamil Nadu Forest Department.

26. Silappathikaram, one of the Five Great Epics of Tamil Literature written by the Poet Elangovadikal.

27. Sivanappan.R.K, Soil and Water Conservation and Water Harvesting, Tamil Nadu Forest Department, 2002.

28. Somasundaram, *A Handbook on the Flora of Southern States*, Government of India, 1963.

29. Sreedharan.C.K. Forest Management Based on Nature and Centered on People, Notion Press, 2020.
30. Sukumar.R et al, A God in distress: threats of poaching and the ivory trade to the Asian elephant in India, 1997.
31. V. Sundararaju, *Maram Manitharin Nanban*, Valvil Printers, 2004.
32. V. Sundararaju, *Management Plan for Kanniyakumari Wildlife Sanctuary*, Tamil Nadu Forest Department, 2007.
33. V.Sundararaju, Jungle Chronicles, Notion Press, 2017.
34. V.Sundararaju, Uyirththidum Ulagam in Tamil, Notion Press, 2017.
35. Tamil Nadu Forest Department, *Forest Management in Tamil Nadu, Past, Present and Future*, 2010.
36. Yoganarasimhan & Chelladurai, *Medicinal Plants of India*, Cyber Media, 2000.

GLOSSARY

Arboreal: living in trees.

Avifauna: the various bird species of a particular region.

Beat: the forest area which is protected by a Forest Guard with the assistance of a Forest Watcher.

Billet: a thick piece of wood.

Biodiversity: biodiversity or biological diversity is the variety or richness of ecosystems, species composition therein and their genetic diversity also.

Breast height: the standard stem height of 1.57 m of a tree from the ground level at which the circumference of a tree is measured is called breast height.

Bull: the male of any animal in the cow family.

Burrow: a hole or a tunnel in the ground made by animals to live in.

Cairn: a mound of rough stones or a structure erected as a boundary pillar or landmark.

Canivore: meat-eater.

Chip: a small piece removed from a hard material such as wood.

Confiscate: seize by the authority to the public treasury.

Covercrop: a crop planted primarily to manage soil fertility, water, weeds, pests, diseases and biodiversity in an agroecosystem.

Crepuscular: deer are crepuscular, meaning that they are most active at dawn and dusk.

Critically Endangered: A taxon is Critically Endangered when it is facing an extremely high risk of extinction in the wild in the immediate future.

Dart: a small pointed syringe used for tranquillizing the animals.

Diurnal: active in the daytime.

Ecosystem: the living community of plants and animals in any area together with the non-living components of the environment such as soil, air and water constitute the ecosystem.

Endangered: a taxon is Endangered when it is not Critically Endangered but is facing a very high risk of extinction in the wild shortly.

Endemic: confined to a particular area.

Extinct: a taxon is Extinct when there is no reasonable doubt that the last individual has died.

Flagship species: the popular species chosen to represent an environmental cause, such as an ecosystem in need of conservation.

Frugivore: fruit eater.

GI: Galvanized Iron.

Girth: the circumference of a tree usually measured at the stem height of 1.57 m from the ground level.

Granivore: grain and seed eater.

Grizzled Giant Squirrel: this squirrel is found in some hill ranges of south India and Ceylon. The dorsal surface and tail have grey or brownish grey, more or less grizzled with white.

Grove: a small wood or group of trees.

Gymnospermous: a vascular plant, such as a cycad or conifer, whose seeds are not enclosed within an ovary.

Habitat: an area that is inhabited by a particular species of animal or plant or other type of organism.

Hatchling: a newly hatched bird, amphibian, fish or reptile.

Herbivore: plant eater.

High forest: bluegum high forest is tall open forest.

Insectivore: plants, animals or birds that feed on insects, worms and other invertebrates.

Log: a bulky piece of a cut or fallen tree.

Mahout: an elephant keeper.

Makhna: tuskless male elephant.

Nectarivore: birds or insects that feed on nectar.

Nilgiris: one of the oldest mountain ranges, located at the tri-junction of Tamil Nadu, Kerala and Karnataka. This is part of the Western Ghats. Nilgiris is India's first biosphere. Because of its unique biodiversity, it has been declared as one of the 14 'Hotspots' of the world.

Nocturnal: active during the night.

Omnivore: animals or birds that eat other animals and plants.

Ootacamund: the district headquarters of the Nilgiris district of Tamil Nadu State in India.

Pachyderm: any thick-skinned mammal, esp. an elephant or rhinoceros.

Predator: an animal preying on others.

Prey: an animal that is hunted or killed by another for food.

Rasam: a syrup of sour taste made with tamarind fruit, tomato, pepper, etc., consumed along with food, especially in South India.

Rattans: any East Indian climbing palm of the genus Calamus with long thin jointed pliable stems.

Salt lick: a salt lick also known as a mineral lick or natural lick is a natural mineral deposit where animals in nutrient-poor ecosystems can obtain essential mineral nutrients.

Sholas: they are patches of stunted evergreen tropical and subtropical moist broadleaf forest found in valleys amid rolling grassland in the higher montane regions of South India. They are located along the Western Ghats in the higher altitude hill regions of Nilgiris and Kanyakumari districts.

SIV Ltd: South India Viscose Ltd, a private wood-based company was set up at Sirumugai, Coimbatore on the bank of the Bhavani River to produce wood pulp from eucalyptus, rayon-grade wood pulp, viscose staple fibre (VSF) and rayon filament yarn.

Subadult: an individual that has passed through the juvenile period but not yet attained typical adult characteristics.

Sylvan: an association with the woods.

TAN India Ltd: a private wood-based company was established to manufacture wattle extract solid and spray-dried powder.

Terrestrial: ground living.

Thicket: a tangle of shrubs or trees.

Threatened: A taxon is Threatened when is likely to become an endangered within a foreseeable future.

Tranquillizer: a drug used to diminish anxiety.

Trellis: a light frame made of long narrow pieces of wood that cross each other.

Troops: especially in large numbers.

Ungulate: a hoofed mammal.

Velvet: the newly grown antler is encased in a thick, soft skin called 'velvet'. Its softness and its dense covering mat of fine hairs give it the feel and the look of velvet. The skin, fed by numerous blood vessels is highly sensitive and easily injured.

Venison: meat from a deer.

Vulnerable: A taxon is Vulnerable when it is not Critically Endangerd or Endangered but is facing a high risk of extinction in the wild in the medium-term future.

Wallowing: large animals lie and roll about in water or mud to keep cool or for pleasure.

Western Ghats: the hill ranges of Western Ghats extend along the west coast of India from the river Tapti of Maharashtra State in the north to Kanyakumari district of Tamil Nadu State, the southern tip of India to a length of nearly 1600 km.

ABBREVIATIONS

ACF	Assistant Conservator of Forests
AD	Assistant Director
AHD	Animal Husbandry Department
AIR	All India Radio
APCCF	Additional Principal Chief Conservator of Forest
APDRP	Andhra Pradesh Disaster Recovery Project
APW	Anti Poaching Watcher
ARC	Arasau Rubber Corporation
AWBI	Animal Welfare Board of India
BBTC	Bombay Burmah Trading Corporation
BCE	Before the Common Era
BMC	Biodiversity Management Committee
BNHS	Bombay Natural History Society
BPL	Below Poverty Line
BRC	Breed Registration Committee
BW	Bungalow Watcher
CAMPA	Compensatory Afforestation Management and Planning Authority
CAG	Comptroller and Auditor General
CAT	Centre for Appropriate Technology

CCF	Chief Conservator of Forests
CE	Common Era
CF	Conservator of Forests
CFRC	Central Forest Rangers College
Ch	Chapter
CLC	Convention on Civil Liability for Oil Pollution Damage
CMFRI	Central Marine Fisheries Research Institute
CR	Conservation Reserve
CRT	Cathode Ray Tube
CRZ	Coastal Regulation Zone
CS	Chief Secretary
CrPC	Criminal Procedure Code
CSWCRTI	Central Soil & Water Conservation Research and Training Institute
CWLW	Chief Wildlife Warden
CWMA	Cauvery Water Management Authority
CZMP	Coastal Zone Management Plan
DAP	Digital Aerial Photogrammetry
DC	District Collector
DCF	Deputy Conservator of Forests
DCTP	Director of Country and Town Planning
DCZMA	District Coastal Zone Management Authority
DCZMC	District Coastal Zone Management Committee
DD	Deputy Director

DDP	Development of Degraded Programme
DE	Divisional Engineer
DFO	District Forest Officer
DPAP	Drought Prone Area Project
DRO	District Revenue Officer
DvlFO	Divisional Forest Officer
DRDA	District Rural Development Agency
DSP	Deputy Superintendent of Police
DWMA	District Water Management Agency
EB	Electricity Board
EIAC	Environmental Impact Assessment Committee
EO	Executive Officer
FC	Forest Contractor
FD	Field Director
FDA	Forest Development Agency
FDC	Forest Development Committee
FG	Forest Guard
FPS	Forest Protection Squad
FRH	Forest Rest House
FRI	Forest Research Institute
FRO	Forest Range Officer
FSO	Fire Service Officer
FSR	Forest Schedule of Rates
FUO	Forest Utilization Officer

FVO	Forest Veterinary Officer
FW	Forest Watcher
GIS	Geographical Information System
GOI	Government of India
GUA	Greening Urban Area
H_2S	Hydrogen Sulphide
HACA	Hill Area Conservation Authority
HADP	Hill Area Development Programme
HPF	Hindustan Photo Films
HR&CE	Hindu Religious & Charitable Endowment
HTCA	Heritage Tree Conservation Authority
HTCC	Heritage Tree Conservation Committee
HTCC	heritage Tree Conservation Centre
HTL	High Tide Line
IAS	Indian Administrative Service
ICAR	Indian Council of Agricultural Research
IFA	Indian Forest Academy
IFS	Indian Forest Service
IGF	Inspector General of Forest
IGP	Inspector General of Police
IIT	Indian Institute of Technology
INTACH	Indian National Trust for Art and Cultural Heritage
IPC	Indian Penal Code
IPS	Indian Police Service

ITDA	Integrated Tribal Development Agency
IUCN	International Union for Conservation of Nature
IWRD	Industrial Wood Research Division
JBIC	Japanese Bank of International Cooperation.
JC	Joint Commissioner
JD	Joint Director
JFMC	Joint Forest Management Committee
JICA	Japanese International Cooperation Agency
JRF	Junior Research Fellow
KFDF	Karnataka Forest Development Fund
KFRI	Kerala Forest Research Institute
KSF	Karnataka State Fund
KSHIP	Karnataka State Highways Improvement Programme
LDA	Local Development Authority
LPG	Liquid Petroleum Gas
LTL	Low Tide Line
LTM	Lion Tailed Macaque
MARPOL	International Convention for the prevention of Pollution from Ships
MD	Managing Director
MFP	Minor Forest Produce
MGNREGA	Mahatma Gandhi National Rural Employment Guarantee Scheme
MLA	Member of the Legislative Assembly

MO	Monitoring Officer
MoEF	Ministry of Environment and Forest
MoEF & CC	Ministry of Environment and Forest & Climate Change
MPCA	Medicinal Plants Conservation Area
NaCl	Sodium Chloride
NAP	National Afforestation Project
NBA	National Biodiversity Authority
NBAGR	National Bureau of Animal Genetic Resources
NBM	National Bamboo Mission
NCCR	National Centre for Coastal Research
NDZ	No Development Zone
NFRC	Northern Forest Rangers College
NGO	Non Governmental Organization
NGT	National Green Tribunal
NMPB	National Medicinal Plants Board
NNFD	Nilgiris North Forest Division
NSFD	Nilgiris South Forest Division
NTFP	Non Timber Forest Product
OA	Office Assistant
OECF	Overseas Economic Co-operation Fund
PA	Personal Assistant
PBR	People's Biodiversity Register
PC	Police Constable
PCB	Polychlorinated Biphenyl

PCCF	Principal Chief Conservator of Forests
PETA	People for the Ethical Treatment of Animals
PO	Project Officer
POB	Prohibitory Order Book
PVC	Polyvinyl Chloride
PW	Plot Watcher
PWD	Public Works Department
RCC	Reinforced Cement Concrete
RDO	Revenue Divisional Officer
RET	Rare Endangered and Threatened
RF	Reserved Forest
RI	Revenue Inspector
RM	Regional Manager
RSP	Road Side Plantation
RSVY	Rashtriya Sam Vikas Yojna
SBB	State Biodiversity Board
SBVNS	Sunset Bazaar Viyabarikal Nala Sangam
SC	Supreme Court
SCO	Soil Conservation Officer
SDMRI	Suganthi Devadasan Marine Research Institute
SDP	Special Development Package
SF	Social Forestry
SFD	Social forestry Division
SFRC	Southern Forest Rangers College

SFSC	State Forest Service College
SI	Sub Inspector of Police
SIV	South India Viscose
SLO	Single Lock Officer
SOFCON	Society for Conservation of Nature
SP	Superintendent of Police
SSFRDT	Society for Social Forestry Research and Development Tamil Nadu
STR	Sathyamangalam Tiger Reserve
TAFCORN	Tamil Nadu Forest Plantation Corporation
TANTEA	Tamil Nadu Tea Plantation Corporation
TAP	Tamil Nadu Afforestation Project
TBGP	Tamil Nadu Biodiversity Conservation & Greening Project
TBCT	Tropical Butterfly Conservatory Tiruchirappalli
TCPL	Tree Cultivation in Private Lands
TFD	Territorial Forest Division
TNAU	Tamil Nadu Agricultural University
TNEB	Tamil Nadu Electricity Board
TNFD	Tamil Nadu Forest Department
TNPPF Act	Tamil Nadu Preservation of Private Forest Act
TWAD	Tamil Nadu Water and Drainage Board
UAV	Unmanned Aerial Vehicle
UK	United Kingdom
UNCLOS	United Nations Convention on the Law of the Sea

UNESCO	United Nations Educational Scientific & Cultural Organization
USA	United States of America
V& AC	Vigilance & Anti-Corruption
VAO	Village Administrative Officer
VFC	Village Forest Committee
WB	West Bengal
WBC	World Bank Consultant
WSHG	Women Self Help Group
WTI	Wildlife Trust of India
WLW	Wildlife Warden

ABOUT THE AUTHOR

Born in an agrarian background in Vairichettipalayam village of Tiruchirappalli District in Tamil Nadu, V.Sundararaju, with Master's degree in Mathematics, has served in Tamil Nadu Forest department for more than 36 years in various capacities such as Forest Range Officer, Assistant Conservator of Forests, Regional Manager, Divisional Forest Officer, Deputy Conservator of Forests and District Forest Officer till he retired from service on 31st January, 2011.

While he was in service, he has published six books in Tamil. Many of his talks and poems were broadcast through the All India Radio of Tirunelveli, Tiruchirappalli, Madurai, Nagercoil and Tuticorin Stations on many occasions. In addition to this, he has delivered hundreds of conservation oriented lectures in different schools, colleges and other public forums.

After retirement, he is giving special talks and lectures on Nature Conservation in various schools, colleges and other public forums. His scientific articles are being published in many of the leading newspapers both in English and Tamil. He is presently serving in the 'Society for Social Forestry Research and Development Tamil Nadu' (SSFRDT) as a Consultant.

He has published one book titled 'Jungle Chronicles' in English and one book with the title 'Uyirththidum Ulagam' in Tamil through Notion Press, Chennai. He is a Blogger in Down to Earth Website. His scientific articles are being published continuously in the website for the past three years. So far forty one articles on various topics like Forests, Wildlife & Biodiversity,

Environment, Water, Agriculture, E-Waste, Mining, Pollution, etc., discussing the scientific strategies for Ecological Harmony have been published in www.downtoearth.org.in.

The present book 'Ecological Harmony' contains the various scientific articles published in 'Down to Earth' website besides notable happenings that had taken place during his service in Tamil Nadu Forest Department. The Author has narrated cogently how he had made use of the opportunities for sustainable management of the Ecological Harmony in the book.

He has authored a book on 'Sacred Trees' on the request made by Rupa Publishers, New Delhi and the same will be published shortly.

He is the President of the forum called 'Society for Conservation of Nature' (SOFCON). Awareness programmes are organised periodically in a college or in any training institute related to Nature and Environment through SOFCON.

ECOLOGICL HARMONY

By
V.Sundararaju

www.ingramcontent.com/pod-product-compliance
Lightning Source LLC
Chambersburg PA
CBHW020731180526
45163CB00001B/191